Materials and Mechanics

LABORATORY EXPERIMENTS

Revised Edition

By Jharna Chaudhuri and Archis Marathe
Texas Tech University, Lubbock

Bassim Hamadeh, CEO and Publisher

Michael Simpson, Vice President of Acquisitions

Jamie Giganti, Managing Editor

Jess Busch, Senior Graphic Designer

Marissa Applegate, Acquisitions Editor

Luiz Ferreira, Licensing Specialist

First published in the United States of America in 2015 by Cognella, Inc.

Printed in the United States of America

ISBN: 978-1-63189-272-1 (pbk) / 978-1-63189-274-5 (sb) / 978-1-63189-273-8 (br)

www.cognella.com   800-200-3908

Contents

Dedication

I dedicate this book to my family (my husband, Bires; my children, Jhuma and Avee; my son-in-law, Andrew James; and my grandchildren, Siona, Kavin, and Rohin James) and to my parents, the late Suresh and Parul Kar.

Jharna Chaudhuri

I dedicate this book to my wife, Neha, and my family.

Archis Marathe

About the Authors

Dr. Jharna Chaudhuri received her PhD from the Mechanics and Materials Department at Rutgers University in 1982. She joined the Mechanical Engineering Department at Wichita State University in 1984, and she was the chair of that department from June 2001 to December 2004. Currently she is the chair of the Department of Mechanical Engineering at Texas Tech University, a position she has held since January 2005. In that capacity, she has moved the department's graduate program and research to a ranking of 81st in the nation from no previous ranking while maintaining an excellent and very large undergraduate program.

Dr. Chaudhuri was a summer faculty research associate at the Wright Patterson Air Force Base in Ohio in 1985 and at the Naval Research Laboratory in Washington, DC, in 1986, 1990, and 1992. She has collaborated with Oak Ridge National Laboratory, Stanford Synchrotron Radiation Laboratory, and Boeing and Cessna in Wichita, Kansas. She has extensive research experience in X-ray diffraction, high-resolution transmission electron microscopy, characterization of metallic and semiconducting materials, synthesis of luminescent nanomaterials for light-emitting diodes and bio applications, and other areas. She has also taught various courses in mechanics and materials.

Archis Marathe is currently pursuing his PhD in mechanical engineering at Texas Tech University. His doctoral research is in the field of nanotechnology. He received a master of science degree in mechanical engineering from Texas Tech University, and a bachelor of engineering degree from University of Pune in India. Archis has been working as an instructor for the mechanics of materials course for more than three years. He is also an electron microscopist and is in charge of the transmission electron microscopy facility at the Department of Mechanical Engineering at Texas Tech University.

Preface

This is the first edition of *Materials and Mechanics Laboratory: Experiments*. This is a two-credit, hour-long course that Professor Chaudhuri has been teaching at Texas Tech University for some time. She felt strongly that students are at a loss without a textbook for this laboratory course. This is our first attempt at creating such a book, and we are hoping that we will improve on it by adding more experiments in the future.

The book is divided into nine chapters, with chapter 1 on the safety protocols we are trying to teach our students, chapter 2 on the procedure for writing lab reports with grading rubrics, and chapters 3 through 9 on seven experiments we are currently teaching in this course. In-class problems are included in each chapters 3 through 9, which will aid better explanation and retention of the studied materials by students. Data sheets are included in each experimental chapter, which students can take to the lab to record data.

We require that 50 percent of reports be group reports and that the other 50 percent be individual reports, both of which should contain all the sections mentioned in chapter 2. This way students can learn how to work in a group, and the same time their individual ability can be assessed.

This book is in no way intended to teach the fundamental and theoretical background of each experiment. There are plenty of materials science engineering and solid mechanics books available that cover the fundamentals. Students should read the relevant chapters from those books. Theory parts in each chapter contain the relevant equations students need for the calculation of different properties in the results section.

PowerPoint presentations and a sample syllabus are available for instructors to use. The authors would also like to extend their help regarding the experimental setup or teaching method as needed (please contact Jharna Chaudhuri (e-mail: Jharna.chaudhuri@ttu.edu) or Archis Marathe (e-mail: archis.marathe@ttu.edu)). This is a consumable book. Students need to bring the data sheet and in-class problem pages to each lab and lecture section, respectively, from the book. This way this is much easier on students and instructors to follow up.

Chapter 1

Laboratory Safety Rules

The purpose of this safety rules is to outline the procedure for providing a safe environment for students in this laboratory or in a similar environment. The policies presented herein are derived from the policies provided by the Environmental Health and Safety, Texas Tech University (please see the reference below).

1. Comply with emergency evacuation procedures.
2. Locate a safety person in each laboratory session (perhaps a student, so that he or she can learn how to handle responsibility) who will be responsible for handling any emergency situation, with supervision of the instructor.
3. Access to eye washes, safety showers, and fire extinguishers must be kept clear.
4. Eating and drinking in the laboratories are **strictly forbidden** at all times!
5. Eye protection (safety glasses or goggles) must be worn at all times in laboratories where chemicals are used and where there is the potential for eye injury.
6. Appropriate gloves (latex, nitrile, butyl, or natural rubber) must be worn at all times when chemicals are used.
7. Lab coats should be worn when dealing with chemicals.
8. Open-toed shoes, sandals, and shorts are not acceptable in laboratories.
9. Keep the chemical hygiene plan, materials safety data sheets (MSDSs), and emergency phone numbers in a highly visible location in the lab.
10. Immediately report any work-related illness or injury to your supervisor and the safety coordinator.
11. Tasks that present unusual hazards must be reviewed with the appropriate supervisor **before** they are conducted.
12. Labels on containers must not be defaced, and all containers of chemicals must be labeled to list the contents, hazards, name of owner, date received, and date opened.
13. Dispose of all expired chemicals per annual inventory of the lab.

14. Gas cylinders must be firmly secured with restraints, whether in use or stored. Regulators must be removed and caps used when moving cylinders.
15. Do not modify electrical equipment yourself! Call the professionals for help.
16. Emergency phone numbers should be posted within the laboratory and on the outside of the laboratory door (911, poison control center, principal investigator's office and home phone numbers, and other employees' or students' home phone numbers).

Reference

Environmental Health and Safety, Texas Tech University, 2903 4th Street, Lubbock, Texas 79409-1090. Copyright 2009–2013 Texas Tech University. All rights reserved.

Chapter 2

Materials and Mechanics Laboratory Report Format

Page 1. Title Page

The title page should include the following:

Course name and number

Title of the experiment

Laboratory section number

Names of group members

Date submitted

Report submitted to: (teaching assistant's name)

Page 2. Summary

This should be the summary of the whole report and should provide a brief description of the objectives, procedure, results, discussion of results, and conclusion. Quantitative results (e.g., percentage of error) should be given to add credibility to conclusions.

Page 3. Table of Contents

The table of contents should include the title of each section and its corresponding page number.

Page 4. List of Figures

Provide a list of figures on a separate page following the table of contents. The figures should be listed in numerical order, with the exact figure captions and the page number where each figure is located.

Page 5. List of Tables

Provide a list of tables on a separate page following the list of figures. Tables must be listed in numerical order, with the exact heading and page number of each table.

Page 6. List of Equations

Provide a list of equations on a separate page following the list of tables. Equations must be listed in numerical order, with the exact description and page number of each equation.

Page 7. The Actual Report Starts from this Page and Should Include the Following:

Introduction

Provide background, applications and objectives justifying the experiment you will be performing:

- Why are you doing this experiment (background)?

- What are the applications?

- What are the objectives of the experiment? Provide a short statement of goals of the experiment. Try to address the following questions for the objectives:

    - What are you trying to do?

    - What kind of results do you expect?

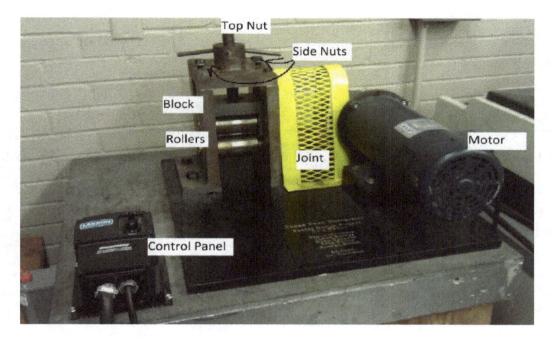

FIGURE 2.1 Cold Rolling Experiment Setup

Theory

Provide the theoretical background of the experiment. This section contains diagrams, sketches, figures, and so forth to describe the proper theoretical background and the complete set of equations you need to use in your laboratory report.

Test Description

The test description section must provide sufficient detail regarding the test setup and samples used. Describe the type of experiment or test, the material tested, and the apparatus used in the experiment or test. Include an illustration of the actual experimental setup used, with proper labeling of each component (see figure 2.1). A detailed description of the apparatus should include make, model, and serial number. Description of the test specimens should include the proper specifications including dimensions, ASTM standards, composition if needed, and images of samples to be used. The test procedure should be briefly described in a paragraph format (preferred).

Results

The results section contains calculated results, graphs, and tables presented in a coherent and understandable manner. Explanations must be given to provide the reader with an understanding of how reduced results were obtained. Each graph, table, and so forth must have a figure caption (full justified) or table heading (centered on the table) and must be referred to in the text in support of the presented results. Figure captions should go at the bottom of the figures. Table headings should appear at the top of the tables. Percentage of difference or percentage of error should be calculated and included in the results section.

Discussion

In this section, the results are discussed through interpretation of data and analysis of errors or differences (i.e., compare your experimental results to published data).

Conclusion

The conclusion should be general and should include a review of the stated purpose of the experiment and discussion of how the experiment satisfied or did not satisfy its objectives. Criticism (positive or negative) and recommendations for improvements should also be included.

References

List all the references you have used in numerical order. Use the proper format for referencing. Follow these examples:

For Citing a Book

[1] Johnson, G. 1985. *Wind Energy Systems*. Englewood Cliffs, NJ: Prentice Hall.

For Citing a Journal Publication or a Conference Proceeding

[2] Sutherland, H., and Veers. P., 1995. "Effects of Cyclic Stress Distribution Models on Fatigue Life Predictions." *Wind Energy* 16:2025.

For Citing a Website

[3] Grand Canyon Skywalk, Wikipedia, the free encyclopedia (2008), http://commons.wikimedia.org/wiki/File:Skywalk_grand_canyon.jpg

Appendix

Include raw data and details of calculations in this section. If you are using Excel, MATLAB, or any other software to do all calculations, include a copy of all of those calculations as well. The first page of the appendix should be the handout data sheet used for the experiment, in which you wrote down all the data taken during the experiment.

Lab Report Writing Format	
Section	**Components**
Title/Cover Page	Course Name and Number Title of Experiment Lab Sec. Name / Names Date Submitted to
Summary	An overview of the entire experiment including: Objectives Procedures Results (including quantitative) & Discussion Conclusion
Lists	Table of Contents List of Figures List of Tables List of Equations
Introduction	Necessary background on the subject Applications Objectives
Theory	Theoretical background Proper Figures & complete set of Equations
Test Setup & Procedure	Complete description of the setup Schematic of set up / image of the set up with proper labeling Specimen description including standards used Complete step by step procedure
Results	Explanations of how reduced results were obtained Results (Tabulated form) Plots
Discussion	Interpretation of data Error analysis Comparison of results to established standard values and discussion
Conclusions	Conclustions reached based on results & discussion Recommendations if appropriate
References	Numerically number list of all used references
Appendix	Datasheets, Sample Calculations, additional figures, charts, etc.

Suggested Scoring Rubric for the Laboratory Report

The structured scoring rubric for the lab report helps instructor to grade lab reports and give feedback to students on how to improve their lab report writing. The instructors are supposed to return graded lab reports before the next lab report is due.

	Section	Points
Report Body	Summary	15
	Introduction	15
	Theory	15
	Test description	15
	Results	20
	Discussion	15
	Conclusion	5
Deductions	Missing title page	−5
	Formatting errors	−5
	Grammatical errors	−5
	Spelling errors	−5
	Missing references	−5
	Missing appendix	−5

Chapter 3

Beam Deflection Experiment

Introduction

In this experiment we will use a cantilever beam that is fixed at one end, with a single load applied at the other end or at any other point along its length.

Cantilever beams have numerous usages in construction, including buildings and bridges. Below are some of the examples of the applications of a cantilever beam. Figure 3.1 is an example where load is distributed at the other end instead of at a single point. Figure 3.2 is another example of a cantilever beam.

Include applications of cantilever beams in your report.

High structural stiffness of the material out of which these beams are made is required. So, as a design engineer, you need to select the stiffest material for these types of beams for applications. You will find in the theory section that the structural thickness depends on the elastic modulus as well as the dimensions of the beam.

The objective of the beam-bending experiment is to determine the structural stiffness of a beam made of different materials (steel and aluminum) but with the same thickness and cross-sectional area. To understand how the structural stiffness of a beam depends on the elastic modulus and dimensions of the beam, and why dimensions are so important in design.

It is expected that with the same dimension, steel would have a higher structural stiffness than would aluminum. It is also expected that to have the same structural stiffness in a beam with the same length and with a lower elastic modulus, higher thickness and width are necessary.

FIGURE 3.1 The Grand Canyon Skywalk, an Example of a Cantilever Beam. Copyright © Purple (CC BY-SA 3.0) at http://commons.wikimedia.org/wiki/File:Skywalk_grand_canyon.jpg.

FIGURE 3.2 The Howrah Bridge in Kolkata, India, Another Example of a Cantilever Beam Copyright © Dilip Muralidaran (CC BY-SA 2.0) at http://en.wikipedia.org/wiki/File:Image-Kolkata_Bridge.jpg.

Theory [1]

The cantilever beam of length l below is supported at one end, point A, and a load, F, is applied at the other end, which causes a deflection of the beam, by δ (see figure 3.3).

At point A the moment, M, is given as

$$M = -Fl. \tag{3.1}$$

The deflection of the beam, δ, is

$$\delta = Fl^3/3EI \text{ in m}, \tag{3.2}$$

where E is the elastic modulus and

$$E = \sigma/\varepsilon \text{ in N/m}^2, \tag{3.3}$$

where σ is the stress and ε is the strain, and I is the area moment of inertia, given as

$$I = bh^3/12 \text{ in m}^4, \tag{3.4}$$

where b = width and h = height of the beam. Substituting for I in equation 3.2, the expression for the deflection becomes

$$\delta = 4Fl^3/Ebh^3. \tag{3.5}$$

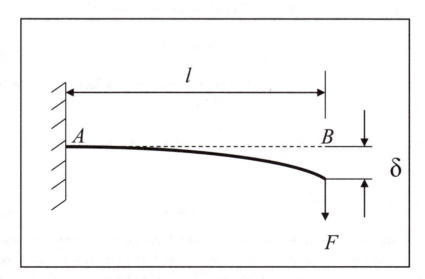

FIGURE 3.3 A Cantilever Beam of Length l, Clamped at One End and Loaded at the Other End

From equation 3.5, the load, F, can be expressed as

$$F = \frac{Ebh^3}{4l^3}\delta \text{ in } N, \qquad (3.6)$$

or, $$F = k\delta, \qquad (3.7)$$

where k is defined as the structural stiffness of the beam and is given as (from equations 3.6 and 3.7)

$$k = \frac{Ebh^3}{4l^3} \text{ in } N/m. \qquad (3.8)$$

Thus the structural stiffness of the beam depends on

- The materials property stiffness or elastic modulus; structural stiffness increases with the elastic modulus of materials.

- More important, structural stiffness depends on the dimensions of the beam; the structural stiffness increases with the width and the height of the beam and decreases with the length.

The deflection of the beam depends on the

- Load linearly; there is more deflection with a larger load.

- Length of the beam; the deflection increases with the length.

- Width and height of the beam; more width and height cause less deflection.

- Material stiffness; higher stiffness produces less deflection.

- Structural stiffness; higher structural stiffness produces less deflection (which is desirable for designing bridges or any other cantilever structure).

It can be seen from equation 3.4 that the deflection is most sensitive to the length and height of the beam.

Test Description [2]

The experimental setup is shown in figure 3.4. Two cantilever beams having the same cross-section geometry (rectangle) and the same dimensions will be used, but one will be made of steel and the other will be made of aluminum. Clamp the beams at one end and measure deflection while you apply known loads. By clamping the beam at one end, you are creating a cantilever beam. Before you start the experiment, take a picture of your test apparatus and samples and include them in your report. Make sure to label every part of the test apparatus. Three experiments are performed as follows:

Experiment 1

Keep the length constant and apply loads at a certain point along the beam (not necessarily the end point; different groups should use different lengths). Use a set of same loads for both aluminum and steel beams and measure deflection each time. Take the deflection measurement at least three times ($i = 1, 2,$ and 3) for each load.

Experiment 2

Keep the load constant and vary the length, and measure deflection each time for both aluminum and steel beams. Take the deflection measurement for each length three times.

Experiment 3

Keep the load constant and vary the height, and measure deflection each time for both aluminum and steel beams. Take the deflection measurement for each height three times.

Test Procedure

- Clamp the first beam at one end and calibrate the deflection meter reading to zero.

- Add the first weight at a suitable length (record the length) making sure that the hanger is stable and the deflection meter needle is touching the beam. Record the deflection reading and write this in the data sheet.

- Repeat this process three times to collect three sets of data.

- Continue to add weights in increasing order (100, 200, 300, 400, and 500 grams), and record the measured deflections at each step.

- Follow the same procedure and collect three sets of data for the second beam.

- In experiment 2, repeat the experiment, keeping the load constant (in grams) and varying the length (100, 200, 300, 400, and 500 mm), and record the deflection reading three times for each length.

- In experiment 3, repeat the experiment, keeping the load constant (in grams) and varying the height of the beam (5, 10, 15, 20, and 25 mm), and record the deflection reading three times for each length.

Materials/Mechanics Laboratory
Beam-Bending Experiment

Experiment 1

Beam-Bending Experiment Data Sheet (Should Be Included in the Report's Appendix)

Aluminum Beam ($E = 69$ GPa):

Dimensions: Length (l) =   Width (b) =   Height (h) =   Moment of Inertia (I) =

Mass (g)	Force (N)	Experimental Deflection (mm)				
		1	2	3	Average	Std. Dev.
100						
200						
300						
400						
500						

Steel Beam ($E = 207$ GPa):

Dimensions: Length (l) =   Width (b) =   Height (h) =   Moment of Inertia (I) =

Mass (g)	Force (N)	Experimental Deflection (mm)				
		1	2	3	Average	Std. Dev.
100						
200						
300						
400						
500						

Materials/Mechanics Laboratory
Beam-Bending Experiment

Experiment 2

Beam-Bending Experiment Data Sheet (Should Be Included in the Report's Appendix)

Aluminum Beam ($E = 69$ GPa):

Dimensions: Width (b) =   Height (h) =   Moment of Inertia (I) =

Mass: 200 g,   Force (F) =

Length (mm)	Length³ (mm³)	Experimental Deflection (mm)				
		1	2	3	Average	Std. Dev.

Steel Beam ($E = 207$ GPa):

Dimensions: Width (b) =   Height (h) =   Moment of Inertia (I) =

Mass: 200 g,   Force (F) =

Length (mm)	Length³ (mm³)	Experimental Deflection (mm)				
		1	2	3	Average	Std. Dev.

Materials/Mechanics Laboratory
Beam-Bending Experiment

Experiment 3

Beam-Bending Experiment Data Sheet (Should Be Included in the Report's Appendix)

Aluminum Beam ($E = 69$ GPa):

Dimensions: Width (b) =   Height (h) =   Moment of Inertia (I) =

Mass: 200 g,   Force (F) =

Height (mm)	Height³ (mm³)	Experimental Deflection (mm)				
		1	2	3	Average	Std. Dev.

Steel Beam ($E = 207$ GPa):

Dimensions: Width (b) =   Height (h) =   Moment of Inertia (I) =

Mass: 200 g,   Force (F) =

Height (mm)	Height³ (mm³)	Experimental Deflection (mm)				
		1	2	3	Average	Std. Dev.

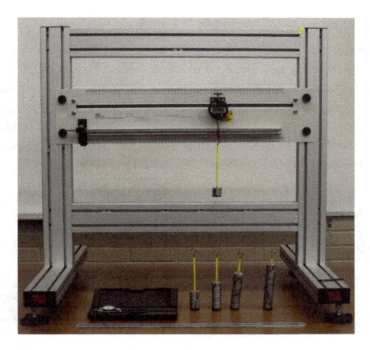

FIGURE 3.4. Test Setup for the Beam-Bending Experiment

Results

For each deflection measurement, x_i, the deviation, d_i, is

$$d_i = x_i - x_{ave},$$ (3.9)

where x_{ave} is the average of three measurements.

The mean deviation, d_{mean}, is (n = total number of measurements = 3 in this case)

$$d_{mean} = \sum_{i=1}^{n} \frac{d_i}{n}.$$ (3.10)

The standard deviation, S, is

$$S = \sqrt{\sum_{i=1}^{n} \frac{(x_i - x_{ave})^2}{(n-1)}}.$$ (3.11)

Experiment 1

Include force and experimental deflection (average values and standard deviation) for each experimental data point from the data sheets in tables 3.1 and 3.2. Calculate theoretical deflection using equation 3.5. Plot load versus experimental and theoretical deflections for both beams in two separate plots.

Find the stiffness, k (slope of the force vs. deflection curve), for each material from the plot of the experimental data point. Calculate theoretical stiffness, k, for each material from the dimension, using the materials stiffness or elastic modulus and equation 3.8. Calculate the percentage of difference. Report these data in table 3.3. Display structural stiffness in figure 3.5.

Table 3.1 Force and Experimental and Theoretical Deflections for the Aluminum Beam

Force (*N*)	Experimental Deflection (mm)		Theoretical Deflection (mm)
	Average	Std. Dev.	

Table 3.2 Force and Experimental and Theoretical Deflections for the Steel Beam

Force (*N*)	Experimental Deflection (mm)		Theoretical Deflection (mm)
	Average	Std. Dev.	

Table 3.3 Theoretical and Experimental Structural Difference and Percentage of Error for Both Beams

Beam	Structural Stiffness (*N/m*)		Percentage of Difference
	Experimental	Theoretical	
Aluminum			
Steel			

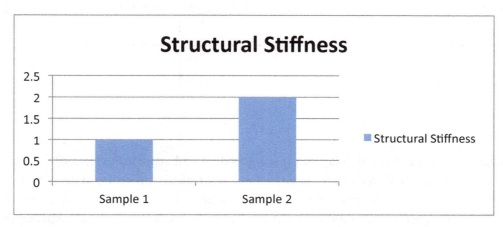

FIGURE 3.5 Example to Display Structural Stiffness of Both Samples in a Column Chart for Immediate and Visual Comparison

Experiment 2

Table 3.4 Length³, Experimental and Theoretical Deflections, and Structural Stiffness for the Aluminum Beam

Length³ (mm³)	Experimental Deflection (mm)		Theoretical Deflection (mm)	Structural Stiffness (N/m)
	Average	Std. Dev.		

Include length³ and experimental deflection (average values and standard deviation) for each experimental data point from the data sheets in tables 3.4 and 3.5. Calculate theoretical deflection and structural stiffness using equations 3.5 and 3.8, respectively, and include these measures in tables 3.4 and 3.5. Plot length³ versus experimental and theoretical deflections for both beams in two separate plots. Plot length³ versus structural stiffness in one graph.

Table 3.5 Length³, Experimental and Theoretical Deflections, and Structural Stiffness for the Steel Beam

Length³ (mm³)	Experimental Deflection (mm)		Theoretical Deflection (mm)	Structural Stiffness (N/m)
	Average	Std. Dev.		

Experiment 3

Table 3.6 Height³, Experimental and Theoretical Deflections, and Structural Stiffness for the Aluminum Beam

Height³ (mm³)	Experimental Deflection (mm)		Theoretical Deflection (mm)	Structural Stiffness (N/m)
	Average	Std. Dev.		

Include height³ and experimental deflection (average values and standard deviation) for each experimental data point from the data sheets in tables 3.6 and 3.7. Calculate theoretical deflection and structural stiffness using equations 3.5 and 3.8, respectively, and include these measures in tables 3.6 and 3.7. Plot height³ against experimental and theoretical deflections for both beams in two separate plots. Plot height³ versus structural stiffness for both materials in one graph.

Table 3.7 Height³, Experimental and Theoretical Deflections, and Structural Stiffness for the Steel Beam

Height³ (mm³)	Experimental Deflection (mm)		Theoretical Deflection (mm)	Structural Stiffness (N/m)
	Average	Std. Dev.		

Discussion

Experiment 1

Discuss all the results obtained. Mention that the deflection is linearly proportional to the applied load and also that the ratio of deflections of steel and aluminum beams is the same as the ratio of their elastic modulus. Discuss which material has higher structural stiffness for the same geometry and give your reasons (refer to figure 3.5).

Compare the theoretical stiffness with the experimental one and discuss the percentage of difference. If there is a discrepancy, explain why this exists.

Discuss the standard deviation.

Experiment 2

Discuss all the results obtained. Mention that the deflection is linearly proportional to $length^3$. Mention that the structural stiffness is inversely proportional to $length^3$.

Discuss the standard deviation.

Experiment 3

Discuss all the results obtained. Mention that the deflection is inversely proportional to $height^3$. Mention that the structural stiffness is directly proportional to $height^3$.

Discuss the standard deviation.

Conclusions

Make your overall conclusions about this experiment. Comment on what you liked about this experiment and what you did not like. Give recommendations for how to improve this experiment.

References

[1] James M. Gere, and Stephen P. Timoshenko., 1997. *Mechanics of Materials*. Boston: PWS Publishing Company.

[2] *Deflections of Beams and Cantilevers Lecturer's Note*. Nottingham, England: TQ Education and Training Ltd, www.tq.com.

Appendix

Present all data sheets and calculations, including hand calculations, in the appendix.

In-Class Problems—Beam Deflection Experiment

1. Calculate the structural stiffness of a cantilever beam with an elastic modulus 200 MPa, length 1 m, width 10 cm, and height 1 cm that is subjected to a 100 N load.

2. What will be the height of a beam made of a different material with an elastic modulus 100 MPa, and of the same length and width, which is subjected to the same load as in problem 1 and has the same structural stiffness?

3. What conclusion do you make after doing these two problems?

Tensile Testing Experiment

Chapter 4

Introduction

In this experiment we will be measuring the mechanical properties of materials using a tensile testing machine. Engineering materials are subjected to a wide variety of mechanical tests to measure their strength, elastic constants, and other mechanical properties. The results of such tests are used for two primary purposes:

- Materials are subjected to loads when in service. Engineering design needs to estimate the load any structure can withstand based on strength and deflection from the elastic modulus.

- Quality control, either by the materials manufacturers or by the end users, to confirm the materials' property specifications.

The objective of this experiment is to determine the mechanical properties of materials, namely, elastic modulus, yield strength, ultimate strength, modulus of resilience, toughness, percentage of elongation, ductility, and percentage of reduction in area. In addition one needs to determine proportional limit, yield point, fracture point, and permanent set from the stress versus strain curve.

The two materials most commonly used are steel and aluminum. It is expected that steel is stronger than aluminum. Heat treatment changes all material properties of metal alloys except the elastic modulus (elastic modulus depends on bonding between two atoms). It is expected that with heat treatment the elastic modulus will remain the same, but strength and ductility will change.

Theory [1, 2]

A standardized sample is used for the tensile test following the American Society of Testing Materials standards (ASTM E8/E8M–11). The specimen has either a round shape (see figure 4.1 [a]) or is flat with a grip area and a gage section. The specimen is held in the grip section and fails in the gage section because this section has smaller cross-sectional area (see figure 4.1 [b]) and hence the stress is this section is higher.

The tensile stress (also called the engineering stress), σ, is the ratio of the tensile load, F, applied to the specimen to its original cross-sectional area, A_0, at the gage section:

$$\sigma = \frac{F}{A_0}. \tag{4.1}$$

The length of the specimen continuously increases under the tensile force until it fails. If the initial gage length of the specimen is l_0, and instantaneous gage length at any time during the experiment is l_i, the strain (also called the engineering strain), ε, is defined as

a

b

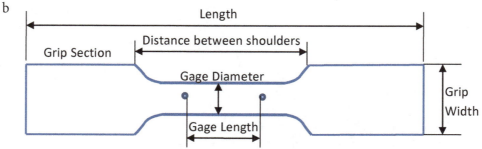

FIGURE 4.1 (a) Tensile Test Samples; (b) Test Specimen Nomenclature

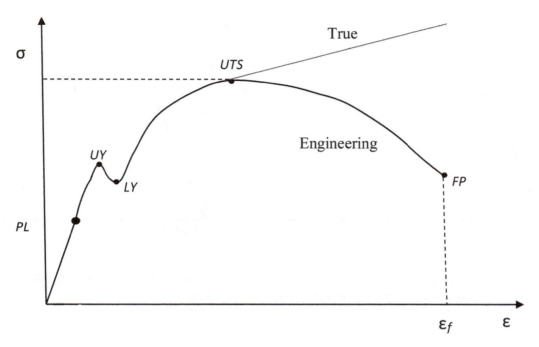

FIGURE 4.2 Typical Stress-Strain Curve of Plain Carbon Steel (*UY* = Upper Yield Strength, *LY* = Lower Yield Strength, *UTS* = Ultimate Tensile Strength, *FP* = Fracture Point, *PL* = Proportional Limit)

$$\varepsilon = \frac{(l - l_0)}{l_0}. \tag{4.1}$$

A typical stress-strain behavior of low carbon steel is shown in figure 4.2. Initially the stress-strain behavior is linear up to the yield point, and after that the behavior becomes nonlinear. The linear portion of the stress-strain curve shows the elastic deformation behavior; that is, if the load is removed, the strain becomes zero (e.g., when you pull a rubber band, it stretches, and when you release it, the band goes back to its original dimension). Nonlinear portion of the stress-strain curve indicates permanent (or plastic) deformation; that is, the strain is permanent and does not become zero when the sample is unloaded (e.g., if you bend a copper tube, you can't make it straight, because it would deform permanently). Thus, there will be a change of the material dimensions remaining after removal of the load. The yield point is when the plastic deformation starts. The corresponding stress is defined as yield strength, σ_y.

Young's modulus, or the modulus of elasticity, E, is defined as

$$E = \frac{\sigma}{\varepsilon}. \tag{4.3}$$

The modulus of elasticity is determined from the initial slope of the stress strain curve as

$$E = \frac{\Delta \sigma}{\Delta \varepsilon}. \tag{4.4}$$

The yield strength, σ_y, is the strength above which plastic deformation occurs:

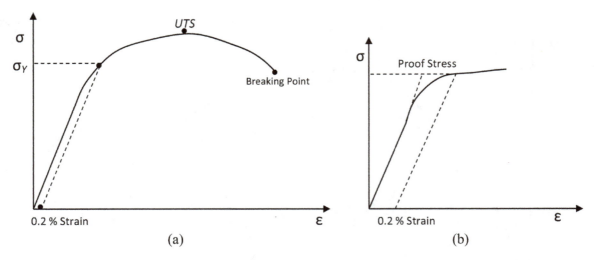

FIGURE 4.3 (a) Stress Versus Strain Curve of a Nonferrous Metal Alloy Such as Aluminum and (b) Method to Determine 0.2 Percent Offset Yield Strength

$$\sigma_y = \frac{F_Y}{A_0},$$ (4.5)

where F_Y is the load at yield point.

The highest stress value before the plastic deformation is called the upper yield strength, and the lower stress value before the plastic deformation is called the lower yield strength. Only ferrous alloys such as low carbon steel show mark yield points.

Nonferrous metals and high carbon steels do not have a definite yield point (see figure 4.3 [a]). The yield strength is chosen as the stress point corresponding to 0.2 percent strain, as shown in figure 4.3 [b]. This stress is also called 0.2 percent offset yield strength.

Ultimate tensile strength, σ_u, is the maximum strength after which reduction in cross-section area in the gage section of the specimen occurs, as shown in figure 4.4. σ_u is determined in the following way:

$$\sigma_u = \frac{F_M}{A_0},$$ (4.6)

where F_M is the load at the maximum point in the stress strain curve.

Resilience is the energy absorbed by the material when it is deformed elastically. Modulus of resilience, U_r, is calculated using the following formula:

$$U_r = \frac{1}{2}\sigma_y\varepsilon_y = \frac{\sigma_y^2}{2E},$$ (4.7)

where ε_y is the strain at the yield strength.

Toughness is the ability of a material's resistance to fracture by absorbing energy and deforming plastically. Toughness is defined as the amount of energy per volume absorbed by a material before fracturing.

Toughness, U_T, can be calculated using the following formula:

$$U_T = \frac{(\sigma_y + \sigma_u)\varepsilon_f}{2}, \tag{4.8}$$

where ε_f is the failure strain.

The ductility of a material is its ability to deform before fracture. Materials with low ductility are brittle, and those with high ductility are ductile. Ductility is measured as the percentage of elongation.

The percentage of elongation is measured as follows:

$$\text{Percentage of elongation} = \left(\frac{l_f - l_0}{l_0}\right) \times 100 = \frac{\Delta l}{l_0} \times 100, \tag{4.9}$$

where l_f is the final length at fracture, and is the change in length in the gage section.

The percentage of reduction in area is another measure of ductility. The percentage of reduction in area can be determined from the following equation:

$$\text{Percentage of reduction in area} = \left(\frac{A_0 - A_f}{A_0}\right) \times 100 = \frac{\Delta A}{A_0} \times 100, \tag{4.10}$$

where A_0 is the initial cross-sectional area, A_f is the final cross-sectional area, and ΔA is the change in cross-sectional area in the gage section.

True stress, σ_t, is obtained by dividing the applied load by the instantaneous cross-sectional area, A_i, rather the initial cross-sectional area, A_0, as follows:

$$\sigma_t = \frac{F}{A_i}. \tag{4.11}$$

True strain, ε_t, is defined as

$$\varepsilon_t = \ln\frac{l_i}{l_0}. \tag{4.12}$$

Figure 4.4 shows the true stress-strain curve. The relation between true stress and strain in the elastic region is as follows:

$$\sigma_t = E\varepsilon_t. \tag{4.13}$$

The relation between true stress and strain in the plastic region is as follows:

$$\sigma_t = K(\varepsilon_t)^m, \tag{4.14}$$

where K is the strength coefficient and m is the strain hardening exponent. The value of m is between 0.1 and 0.5 and gives a measure of strain hardening in materials. A higher value of m means that the

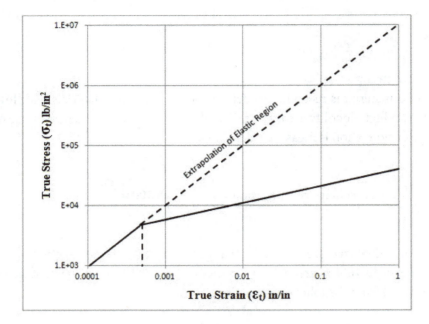

FIGURE 4.4 Plot of True Stress Versus True Strain in a Logarithmic Scale

material is more strain hardened or stronger. Taking the logarithm of both sides of equation 4.14, it can be written as

$$\log \sigma_t = \log K + m \log \varepsilon_t.$$ (4.15)

The slope of the plot of σ_t against ε_t in the plastic region gives the value of m, and the intersection of the extrapolated line to zero true strain gives the value of K (see figure 4.4).

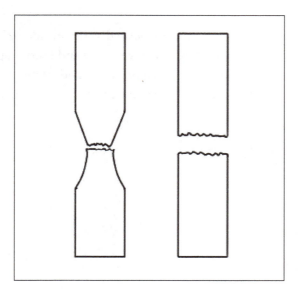

FIGURE 4.5 Schematics of Ductile (Left) and Brittle (Right) Fractures

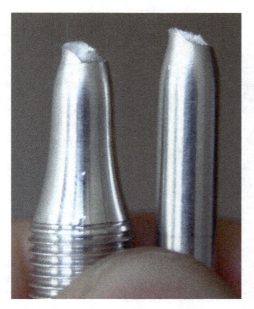

FIGURE 4.6 (left) Cup and Cone Ductile Fracture in Aluminum; (right) Flat Brittle Fracture in Mild Steel.
Copyright © Sigmund (CC BY-SA 3.0) at http://commons.wikimedia.org/wiki/ File:AL_tensile_test.jpg.
Copyright © Sigmund (CC BY-SA 3.0) at http://commons.wikimedia.org/wiki/ File:Cast_iron_tensile_test.JPG.

Fractures that occur from uniaxial tensile load can be ductile or brittle. In ductile metal alloys, necking, or reduction of cross-sectional area, occurs in the gage section right after the maximum point at the stress-strain curve, σ_u. Necking continues until the specimen fractures. If the specimen breaks by necking, the fracture surface will have a cup and a cone shape, which is typical of the ductile fracture (see figures 4.5 and 4.6 [a]). Ductile fracture is associated with considerable plastic deformation and energy absorption. The ductile fracture surface will have a gray and fibrous appearance.

Brittle fracture is associated with very little plastic deformation and energy absorption, and it will result a flat fracture surface. The brittle fracture surface will have a shiny, granular appearance (see figures 4.5 and 4.6 [b]).

FIGURE 4.7 Materials Testing Systems (MTS) Tensile Testing Machine

Test Description [1, 2]

Figure 4.7 shows the MTS tensile testing equipment. The American Society for Testing and Materials (ASTM) standardized test method is used to prepare the test specimen, conduct the test, and analyze the data.

Take pictures of all the equipment used for this experiment and include them in your lab report. Label all components. Write the model number of the equipment.

Give specifications of the test specimens, and include relevant data. Four different samples to be tested are as follows:

AISI 1045, AISI 1018, Aluminum 2024-0, Aluminum 2024-T6

Document the composition and heat treatment of different samples you will be analyzing (make a table). Take a picture of the specimens, and include it in your lab report.

Measure the diameter and length of the sample in the gage section using a slide caliper before and after the experiment (after the samples fail).

Test Procedure

Use the following procedure to run the test:

- Turn on the hydraulics.

- Start the actuator on low pressure.

- Adjust the bottom grip to the desired position.

- Adjust the gripping pressure to an appropriate value.

- Grip the test specimen.

- Perform the load calibration: Load à Setupà Calà Autoà Go!

- Open the Wavedit program.

- Check and set the memory storage.

- Check and adjust the waveform parameters.

- Press RUN.

- In the Runtime window, press RUN again.

- Open the file and then select OK.

- Select Low Pressure and then select OK.

- Ignore the message indicating All Channels not Calibrated.

- Press the remote button on the control panel when prompted.

- Press OK to start the test.

- After the specimen breaks, press the STOP button in the Runtime window.

- Close the Runtime window.

- After the Function inhibitor … error message appears, press Resume.

- Turn off Load Protect.

- Remove the broken specimen.

- Either repeat the procedure from gripping the test specimen for a new test, or turn the hydraulics OFF.

Table 4.1 Initial and Final Gage Diameter, Gage Length and Cross-Sectional Area of Four Samples

Sample	Gage Diameter (cm)		Gage Length (cm)		Cross-Sectional Area (cm²)	
	Initial (d_0)	Final (d_f)	Initial (l_0)	Final (l_f)	Initial (A_0)	Final (A_f)

Table 4.2 Experimental Mechanical Properties of Four Samples

Sample	Elastic Modulus (GPa)	Yield Strength (MPa)	Ultimate Strength (MPa)	Modulus of Resilience (Pa)	Toughness (MPa)	Percentage of Elongation	Percentage of Reduction in Area

Table 4.3 Reference Elastic Modulus, Yield Strength, Ultimate Strength, and Percentages of Difference

Elastic Modulus (GPa) (Reference)	% Error	Yield Strength (MPa) (Reference)	% Error	Ultimate Strength (MPa) (Reference)	% Error

Table 4.4 Strength Coefficient (Experimental) and Strain Hardening Exponent (Experimental, Reference, and Percentage of Difference) of Four Samples

Sample	Strength Coefficient, K (MPa)	Strain Hardening Exponent, m		
		Exp.	Ref.	% Diff.

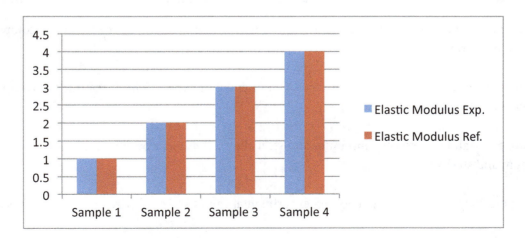

FIGURE 4.8. Example to Display Modulus and Yield and Ultimate Strength of All Four Samples in a Four-Column Chart for Immediate and Visual Comparison

Pull the sample to fracture it. During this step, the crosshead (i.e., the upper part of the machine) remains stationary while the actuator moves down. Stored on the computer are actuator positions and reaction forces as determined by a load cell, which is a device that measures force.

Retrieve the force and displacement data from the computer. Transform force to stress and displacement to strain. This set of data goes to the appendix.

Take a picture of the sample when necking starts, and also take a picture of all fracture surfaces; include these pictures in your report.

Results

Document gage diameter and length for each sample in table 4.1. Calculate cross-sectional area and document this in table 4.1.

Calculate engineering stress by dividing the load by the initial cross-sectional area, and calculate engineering strain from the displacement and initial gage length.

Plot the engineering stress-strain curve. Calculate the elastic modulus (E), yield strength (σ_y), ultimate strength (σ_u), modulus of resilience (U_r), toughness (U_T), percentage of elongation, and percentage of reduction in area for all four samples, and include these measures in table 4.2.

Compare the experimentally obtained data of E, σ_y, and σ_u of all four samples with published or reference data (you should be able to get the published data from the Internet). Calculate the percentage of error. Include in Table 4.4.

Make column charts of E, σ_y, and σ_u of all four samples, as shown in figure 4.8.

Calculate the instantaneous cross-sectional area by dividing the volume at the gage section by the instantaneous length. Calculate the true stress and strain. Plot true stress versus true strain in a log-log graph, and find k and m. Include these values in table 4.4.

Get published data for m. Compare experimentally obtained m values with the published data and include in table 4.4.

Document whether necking occurred before fracture and the nature of the fractured surfaces.

Discussion

Compare E, σ_y, and σ_u of the two steel alloys. Comment on the effect of heat treatment or composition on mechanical properties.

Compare E, σ_y, and σ_u of two aluminum alloys. Comment on the effect of heat treatment or composition on mechanical properties.

Compare E, σ_y, and σ_u between aluminum and steel alloys. Comment on the mechanical properties of aluminum and steel alloy.

Discuss percentages of differences between experimental values and published data for all four samples.

Compare K and m for all four samples.

Discuss all fracture surfaces (ductile or brittle fracture, necking, etc.).

Conclusions

Make your overall conclusions about this experiment. Comment on what you liked about this experiment and what you did not like. Give recommendations for how to improve this experiment.

References

[1] 2013. *ASTM Standard E 8, Standard Test Methods for Tension Testing of Metallic Materials.* West Conshohocken, PA: ASTM International.

[2] 2000. *ASM Handbook Vol. 8, Mechanical Testing and Evaluation.* Materials Park, OH: ASM International.

Appendix

Present load displacement data for all four samples and calculations, including hand calculations, in the appendix.

Note to Instructors

If time permits to add one more lab experiment, this experiment can be extended to another lab experiment on different types of composite materials, where specific modulus (modulus/density) and specific strength (strength/density) values can be compared.

Name: **Lab Section:**

In-Class Problems—Tensile Testing Experiment

Show all calculations. For in-class problems 1 through 8, refer to the following figure:

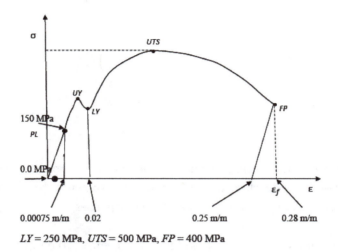

$LY = 250$ MPa, $UTS = 500$ MPa, $FP = 400$ MPa

Answer the following questions:

1. What test is represented by the diagram?

2. What is the material that was tested to produce these results?

3. What is the modulus of elasticity?

4. What is the percent elongation at failure?

5. What is the yield strength?

6. What is the ultimate strength?

7. What is the Modulus of resilience?

8. What is the toughness?

9. A cylindrical specimen of magnesium having a diameter of 12 mm and a gage length of 30 mm is pulled in tension. Use the load-elongation characteristics shown in the following table to plot the stress versus strain curve, and compute the following properties:

(A) yield strength, (B) elastic modulus, (C) modulus of resilience, (D) ultimate strength, (E) percentage elongation at failure, and (F) toughness.

Load (N)	Length (mm)
0	30
5,000	30.0296
10,000	30.0592
15,000	30.0888
20,000	30.15
25,000	30.51
26,500	30.90
27,000	31.50
26,500	32.10
25,000	32.79
Fracture	

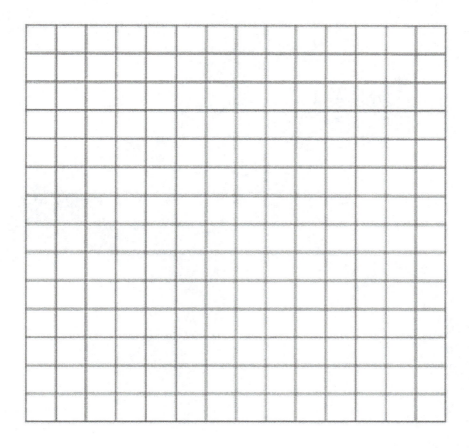

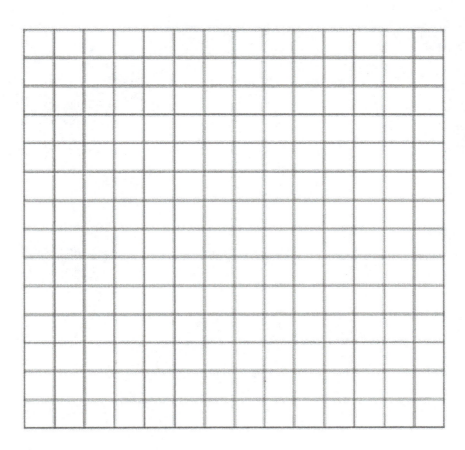

Chapter 5

Hardness Testing Experiment

Introduction

Hardness is the ability of materials to resist plastic deformation or permanent indentation. All metal parts are cold-processed, with rolling, extrusion, stamping, and so forth at room temperature to give a permanent shape. Processing causes generation and entanglement of dislocations, which increase stress and makes metals brittle. Heat treatment after processing is necessary to relieve stress. The hardness test is used to verify the mechanical properties of the materials after the heat treatment and to determine whether a material has the properties necessary for its intended use. In addition, the tensile strength can be estimated from hardness values.

Hardness numbers are also needed for those applications where surface wear is a problem, as in bearings and gears (e.g., airplane landing gears, cutting tools, working tools, etc.).

The objective of this experiment is to determine the hardness of steel and aluminum subjected to various processing and/or heat treatment.

It is expected that steel will have a higher hardness than that of aluminum. Processing such as rolling increases hardness, and so on.

Theory

In the hardness test, an indenter made of harder materials than the materials to be tested is used. Most of the indenters are made of either diamond or tungsten carbide. The shape of the indenter is a ball, a pyramid, or a cone. A load is applied, and the indenter is pressed on the metal surface to make a permanent indentation. The hardness value is calculated based on the dimensions of the indenter and indentation and the amount of load applied.

Three different hardness values will be mentioned here. These are Rockwell, Brinell, and Vickers hardness, as described below.

(a)   (b)

FIGURE 5.1 Rockwell Hardness (a) Testing Equipment and (b) Indenters Used

Rockwell Hardness Test [1]

In the Rockwell hardness test (see figure 5.1), either a diamond cone or a steel ball is used. A minor load of 10 kg is first applied, followed by a major load of 60, 100, or 150 kg. A scale symbol of $A – K$ is used, depending on the indenter and major load applied, as shown in table 5.1.

Table 5.1 Rockwell Hardness Scales and Applications

Scale	Indenter	Major Load (kgf)	Application
A	Diamond cone	60	Cemented carbides, thin and shallow case-hardened steels
B	1/16 in. steel ball	100	Copper and aluminum alloys, soft steels, malleable iron, etc.
C	Diamond cone	150	Cast iron, steel, case-hardened steel
D	Diamond cone	100	Thin and medium case-hardened steel, pearlitic malleable iron
E	1/8 in. steel ball	100	Cast iron, aluminum, and magnesium alloys
F	1/16 in. steel ball	60	Annealed copper alloys, thin soft sheet metals
G	1/16 in. steel ball	150	Phosphor bronze, beryllium copper, malleable iron, aluminum, zinc, lead
H	1/8 in. steel ball	60	
K	1/8 in. steel ball	150	

Each of these scales is intended to be used for specific materials. The hardness value is automatically displayed in the Rockwell hardness tester. For example, 80 HRB means Rockwell hardness 80 in B scale. If the hardness value is greater than 100 or lower than 20, the next scale should be used. In case of a round-shaped specimen, a correction factor has to be added to the Rockwell hardness number.

Brinell Hardness Test [2]

In the Brinell hardness test (see figure 5.2), a 10 mm diameter steel or a tungsten carbide ball is used as an indenter, with a load ranging between 500 to 3000 kg in an increment of 500 kg. The Brinell hardness number, BHN, is obtained as follows:

$$\text{BHN} = \frac{F}{\frac{\pi}{2}D(D - \sqrt{D^2 - D_i^2})}, \tag{5.1}$$

where F is the applied load, D is the diameter of the indenter (and is 10 mm), and D_i is the diameter of indentation.

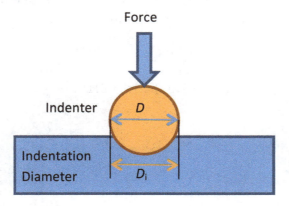

FIGURE 5.2 Brinell Hardness Testing Technique: Indenter and Indentation

Vickers Micro Hardness Test [3]

In the Vickers micro hardness test (see figure 5.3), a small pyramid-shaped diamond indenter is used, with an applied load only between 1 and 100 kg. The Vickers micro hardness number, HV, is calculated as follows:

$$HV = 1.854 \frac{F}{d^2},$$ (5.2)

where F is the applied load, and d is obtained as

$$d = \frac{(d_1^2 + d_2^2)}{2},$$ (5.3)

where d_1 and d_2 are the dimensions of the indentation, as shown in figure 5.3.

Relation between Hardness and Tensile Strength

The tensile strength (TS) increases as the hardness increases. A rough estimate of the tensile strength can be obtained from the Brinell hardness number, as follows:

$$TS\,(MPa) = 3.45 \times BHN$$ (5.4)

$$TS\,(psi) = 500 \times BHN$$ (5.5)

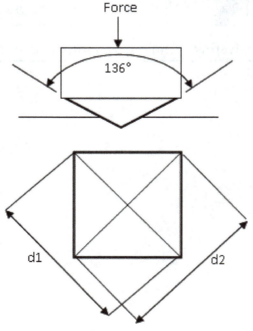

FIGURE 5.3 Vickers Micro Hardness Testing Technique: Indenter and Indentation

Test Description

You will be measuring the hardness of four different samples as listed in the data sheet using the appropriate Rockwell hardness scale. Take a picture of your test apparatus and samples, and include this in your report. Make sure to label every part of the test apparatus.

Test Procedure

- Turn on the Rockwell tester.

- Insert the appropriate indenter and anvil.

- Select the appropriate indenter type on the control panel.

- Select the appropriate load.

- Place the specimen on the anvil.

- Raise the anvil slowly by rotating the handle below it to apply preload. The display dial must read at least 360 (but NEVER more than 360.8).

- Press the start button to start the test.

- Record the hardness value.

- Take three measurements, and show the standard deviation (S).

- Lower the anvil, replace the specimen, and repeat the procedure.

Materials/Mechanics Laboratory
Hardness Testing Experiment

Hardness Test Data Sheet (Should Be Included in the Report's Appendix)

Specimen and Rockwell Scale	Specimen Composition	HR Reading #1	HR Reading #2	HR Reading #3	Ave.	Std. Dev.	BHN	Tensile Strength
Aluminum 2024-O (Flat) HRB								
Aluminum 2024-T6 (Round) HRB								
AISI 1090 Steel (Round) HRC								
AISI 1018 Steel (Flat) HRB								

Results

Get the mean and standard deviation of HR hardness data, and report them in table 5.1. For each Rockwell hardness measurement, x_i, the deviation, d_i, is

$$d_i = x_i - x_{\text{ave}},\tag{5.6}$$

where x_{ave} is the average of three measurements.

The mean deviation, d_{mean}, is (n = total number of measurements = 3 in this case)

$$d_{\text{mean}} = \sum_{i=1}^{n} \frac{d_i}{n}.\tag{5.7}$$

The standard deviation, S, is

$$S = \sqrt{\sum_{i=1}^{n} \frac{(x_i - x_{\text{ave}})^2}{(n-1)}}.\tag{5.8}$$

Table 5.2 Rockwell Hardness Test Results with Mean and Standard Deviation and Reference or Published Values, BHN Experimental and Reference Values, and Tensile Strength Experimental and Reference Values

Speci-men	Comp-osition	Ave. HR EXP.	Std. Dev.	HR REF.	Percen-tage of Error	BHN EXP.	BHN REF.	Percen-tage of Error	Tensile Strength EXP.	Tensile Strength REF.	Percen-tage of Error

Correct for the round shape of the sample if needed using the appropriate table. Obtain the corresponding Brinell hardness values from a table. Obtain the corresponding tensile strength from a table or calculate using equation 5.4.

Get published data for Rockwell hardness, Brinell hardness, and tensile strength for all four samples. Calculate the percentage of difference. Include all data in table 5.2.

Make three separate column charts to show experimental and published, or reference, values for HR hardness, Brinell hardness, and tensile strength values, as in the example shown in figure 5.4.

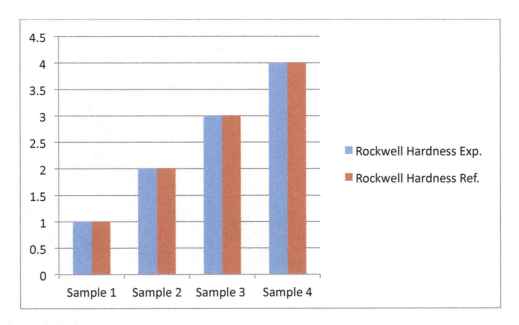

FIGURE 5. 4 Example to Display Hardness (HR and BHN) and Tensile Strength of All Four Samples in a Four-Column Chart for Immediate and Visual Comparison

Discussion

Compare the experimental Rockwell and Brinell hardness and tensile strength with published data. Explain any discrepancies.

Compare the hardness and tensile strength of as-received and tempered aluminum, of 1018 and 1090 steel, and of aluminum and steel.

Conclusions

Make your overall conclusions about this experiment. Comment on whether the objective of the experiment is satisfied or not. Give your recommendations for how to improve this experiment.

References

[1] *ASTM Standard E 18, Standard Test Method for Rockwell Hardness of Metallic Materials*. West Conshohocken, PA: ASTM International.

[2] *ASTM Standard E 10, Standard Test Method for Brinell Hardness of Metallic Materials*. West Conshohocken, PA: ASTM International.

[3] *ASTM Standard E 92, Standard Test Method for Vickers Hardness of Metallic Materials*. West Conshohocken, PA: ASTM International.

Appendix

Present all data sheets and calculations, including hand calculations, in the appendix.

Name: **Lab Section:**

In-Class Problems—Hardness Testing Experiment

1. The Rockwell hardness of a material is HRC 50. Using the appropriate table(s), find the corresponding Brinell hardness and Vickers hardness numbers and tensile strength.

2. A Brinell hardness of 75 is obtained using a 10 mm diameter hardened steel with a 500 kgf load applied. What is the indentation diameter D_i?

3. A Vickers hardness of 800 HV is obtained using a 10 kgf force. What is the corresponding d value?

Chapter 6

Impact Testing Experiment

Introduction

Toughness is a measure of the amount of energy a material can absorb before it fractures. A high toughness value means that the material is ductile, while a low toughness value indicates brittleness of materials. Toughness can be measured using an impact testing machine, in which it is measured under dynamic loading.

During World War II, cargo ships called liberty ships built in the United States were having a problem of brittle fractures initiated not by welding but because of their being subjected to a cold temperature in the North Atlantic Ocean, since the mechanism of failure changed from ductile to brittle with a change of temperature (see figure 6.1) [1].

The resistance of the vehicle in the case of collision is, naturally, a major concern in the car industry. The behavior of welded structures during dynamic or high strain rate loading, reproducing the conditions of collision, is a major concern in automobile industry. Thus, the impact testing of welded thin sheets submitted to the dynamic loading is important.

Impact testing in polymeric material is also very important. Brittle failure of polymers used in booster rocket O-rings and other factors led to the *Challenger* disaster.

The objective of the impact testing experiment is to evaluate the energy-absorbing characteristics of ferrous metals at room temperature.

It is expected that ferrous metals, especially low-carbon steels, go through a ductile brittle transition, being brittle at low temperatures and ductile at high temperatures.

Theory [2]

The transition temperature at which, under impact conditions, material's behavior changes from ductile to brittle is defined as ductile brittle transition temperature (DBTT). This change in the behavior

FIGURE 6.1 One of Only Two Liberty Ships that Are Still Operational. The *SS Jeremiah O'Brien* Was Built in Just Fifty-Five Days and Was Present at the Normandy Invasion.
Copyright © Rennett Stowe (CC BY 2.0) at http://commons.wikimedia.org/wiki/File:World_War_II_Liberty_Ship_%285839499171%29.jpg.

is affected by many variables. Metals that have a face-centered cubic crystalline structure, such as aluminum and copper, have many slip systems and are the most resistant to low energy fracture at a low temperature, and therefore don't go through a ductile brittle transition behavior. Most metals with a body-centered cubic structure (e.g., steel) or a hexagonal crystal structure show a sharp transition temperature and are brittle at low temperatures. Considering steel, coarse grain size, strain hardening, and certain minor impurities can raise the transition temperature, whereas fine grain size and certain alloying elements will increase the low temperature toughness.

Figure 6.2 shows the specimen and standard configurations for the Charpy impact test. The Charpy impact machine consists of a rigid specimen holder and a swinging pendulum hammer for striking the impact blow (see figure 6.3). The pendulum hammer is released from a known height to strike the sample to fracture and bounces to a certain height after the impact (see figure 6.4). The energy imparted to the specimen to fracture is calculated in the following way:

If h_1 and h_2 are the initial and final height of the hammer, respectively, the impact energy absorbed by the specimen to fracture is given as follows:

$$\text{Energy} = mgh_1 - mgh_2, \tag{6.1}$$

where m is the mass of the hammer and g is gravitational acceleration.

The machine is calibrated to read the fracture energy directly from a pointer, which indicates the angular rotation of the pendulum after the specimen has been fractured. The test is performed on samples subjected to five different temperatures. The data is plotted, and DBTT is determined.

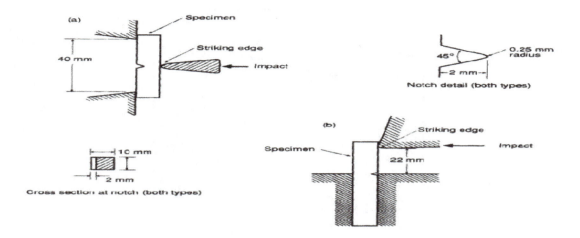

FIGURE 6.2 Specimen and Loading Configuration for Charpy Impact Testing (per ASTM E 23)
Source: Standard Test Methods for Notched Bar Impact Testing of Metallic Materials (ASTM E23). Copyright © by American Society for Testing & Materials (ASTM). Reprinted with permission.

Several different definitions are used for the transition temperature. We define it as the temperature for which the impact energy is halfway between the impact energy for the highest temperature and the impact energy for the lowest temperature. This definition can be expressed as follows:

FIGURE 6.3 Charpy Impact Testing Machine.
Copyright © Cjp24 (CC BY-SA 3.0) at http://commons.wikimedia.org/wiki/File:Charpy_impact_tester.jpg.

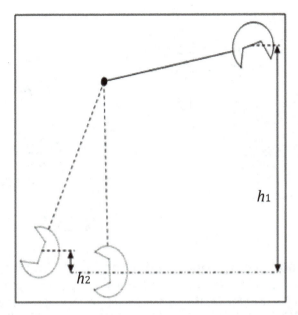

FIGURE 6.4 Schematic Drawing of the Pendulum Hammer Position before and after the Impact

$$\text{DBTT} = \frac{T_t + T_b}{2}, \tag{6.2}$$

where T_t and T_b are the temperatures at the ductile and brittle region, respectively.

Test Description [2]

Test materials to be used are 1045 and 1018 steel. Take a picture of the equipment and test specimen, and include this in your report with proper labeling. Measure the sample dimensions and notch dimension (mention the ASTM standard used to make the specimens). Include the dimensions and composition of sample materials in a table in the results section. The following are the test temperatures to be used:

- Liquid nitrogen: −195.8°C

- Dry ice: −78.6°C

- Plain ice: 0°C

- Room temperature: 25°C

- Boiling water: 100°C

Test Procedure

- Safety is most important. Safety glasses, closed shoes, and long pants should be worn by all participants.

- Pull the pendulum arm back and engage the safety latch, prohibiting movement of the pendulum arm.

- Place the specimen in the correct position.

- Raise the pendulum arm to the test position and firmly support it with the latching mechanism.

- Release the pendulum by pushing up on the release knob. The hammer will drop and attain a striking velocity, striking the specimen, with a swing through a direct reading of the energy absorbed by the specimen. Read the energy absorbed by the specimen and record it.

- Apply the brake until the pendulum has returned to its stable hanging vertical position.

- Remove the specimen from the testing area.

- Leave the pendulum in the down hanging vertical position until another test is to be performed.

Take a picture of all fracture surfaces and include this in your report.

Materials/Mechanics Laboratory
Impact Testing Experiment

Impact Testing Data Sheet (Should Be Included in the Report's Appendix)

Temperature (°C)	Impact Energy Reading (Ft-Lbs)	

Results

Plot the absorbed energy versus temperature, and obtain DBTT for both materials. Calculate the percentage of difference from the published data and include in Table 6.1.

Plot DBTT versus carbon composition of the two materials.

Include pictures of fracture surfaces.

Table 6.1 Experimental and Published DBTT Values and Percentage of Difference

| Materials | Carbon Composition | DBTT | | % Diff. |
		Exp.	Publ.	

Discussion

Compare published DBTT with the experimental DBTT and comment on this. If there is a discrepancy, explain why this exists.

Discuss the dependence of the DBTT on the carbon composition. (The DBTT should decrease with the increase in carbon composition.)

Discuss the nature of the fracture surfaces (ductile and brittle) and any correlation between the fractures surface, test temperature, and DBTT.

Conclusions

Make your overall conclusions about this experiment. Comment on what you liked about this experiment and what you did not like. Give recommendations for how to improve this experiment.

References

[1] Sawyer, L. A., and Mitchell, W. H.. 1985. *The Liberty Ships,* 2nd ed. London: Lloyd's of London Press Ltd.

[2] 1982. *ASTM Standard E 23, Notched Bar Impact Testing for Metallic Materials.* West Conshohocken, PA: ASTM International.

Appendix

Present all data sheets and calculations, including hand calculations, in the appendix.

Name: **Lab Section:**

In-Class Problems—Impact Testing Experiment

1. Define DBTT.

2. A pendulum hammer weighing 10 lb. force strikes a sample to fracture from an initial height of 3 ft. and swings to a final height of 1 ft. What is the impact energy of the sample?

3. The following data were obtained from impact tests performed on a ductile cast iron. Plot the data, and determine the transition temperature (DBTT) from the graph.

4. Would you use this cast iron at −25°C? (Give your reasons.)

Test Temperature (°C)	Impact Energy (J)
−75	2.5
−50	2.5
−25	3
0	6
25	13
50	17
75	19
100	19
125	19

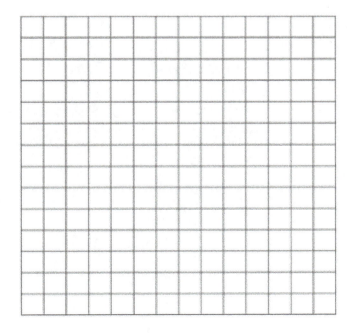

Chapter 7

Heat Treatment and Qualitative Metallographic Analysis (HTQMA)

Introduction

Every mechanical engineer should learn how to do heat treatment of metals and metal alloys in order to come up with the best materials for their design applications.

Plain-carbon steels (in the iron-carbon system) are some of the most important and widely used metals because of their low cost and good properties. Plain-carbon steels are classified into many types according to their composition, and each type has its unique characteristics.

The objective of this metallurgical experiment is to find how the hardness and microstructure change with heat treatment of steels.

FIGURE 7.1 Application of Plain Carbon Steels
Copyright © Jared C. Benedict (CC BY-SA 3.0) at http://commons.wikimedia.org/wiki/File:Gears_large.jpg.

It is expected that the as-received sample will have higher hardness than the annealed one will. The as-received sample is cold-worked and will have elongated grains, and the annealed sample will have larger grains. The amount and distribution of microstructures in both samples will probably be the same. The quenched sample will have much higher hardness and a different microstructure.

Theory [1]

In order to understand the microstructure, you need to review the phase diagram. Phase diagrams are like a road map or a GPS system for heat treatment. A phase diagram is a graphical representation of phases present in a materials system (having more than one component) at various temperatures and compositions. To start with, we like to explain the microstructure of a chocolate-vanilla system (a hypothetical phase diagram), as shown in figure 7.2.

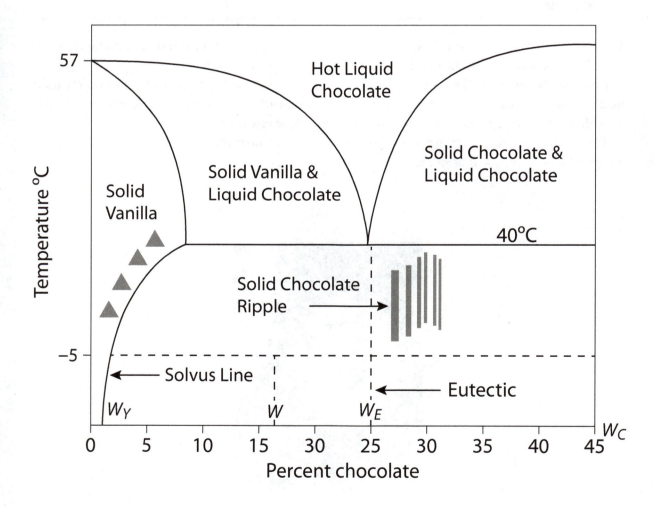

FIGURE 7.2 A Hypothetical Phase Diagram of a Chocolate-Vanilla System

Different Phases and Microstructure Present in the Vanilla-Chocolate System

If you check the vanilla-chocolate system under an optical microscope, you will find the following microstructures in different regions in the phase diagram:

- Liquid in the hot chocolate region.

- Liquid plus solid chocolate malt in the chocolate malt and chocolate region (we would love that!).

- Liquid plus solid vanilla malt in the vanilla malt and chocolate region.

- Solid vanilla malt in the vanilla region.

- Alternate layers of solid vanilla and solid chocolate malt (vanilla-chocolate ripple, we would love that too!) at the eutectic line, as shown in figure 7.3.

- Mostly solid vanilla malt (white in color) and some vanilla-chocolate ripple at the left side of the eutectic line.

- Mostly solid chocolate malt and some vanilla-chocolate ripple at right side of the eutectic line.

FIGURE 7.3 Ice Cream with a fudge ripple

The weight percentages or fractions of phases in any two phase regions can be calculated by using the lever rule. For example, for a vanilla-chocolate ice cream with W wt. percent chocolate at −5°C, a phase analysis using the lever rule would provide the weight percentages of solid vanilla malt and solid chocolate malt as follows:

$$\text{Weight percentage of the solid vanilla malt} = \frac{W_C - W}{W_C - W_V} \times 100, \tag{7.1}$$

and

$$\text{Weight percentage of the solid chocolate malt} = \frac{W - W_V}{W_C - W_V} \times 100, \tag{7.2}$$

where W_V and W_C are the compositions of vanilla and chocolate in that ice cream, respectively.

Similarly, a microstructure analysis can be performed using the same lever rule. Thus, the fractions of solid vanilla malt and vanilla-chocolate ripple present are as follows:

$$\text{Weight percentage of the solid vanilla malt} = \frac{W_E - W}{W_E - W_V} \times 100, \tag{7.3}$$

and

$$\text{Weight percentage of the vanilla} - \text{chocolate ripple} = \frac{W - W_V}{W_E - W_V} \times 100, \tag{7.4}$$

where W_E is the eutectic composition.

Different Phases and Microstructure Present in the Iron-Carbon System

Figure 7.4 shows the phase diagram of the iron-carbon system. Microstructures of interest are as follows:

- Solid Austenite grains in the austenite solid solution region.

- A plain carbon steel with approximately 0.8 wt. percent carbon or at the eutectoid composition and at a temperature of 723°C or lower has the pearlite microstructure, which is a lamellar structure with alternate layers of ferrite (α iron phase, white color) and cementite (Fe$_3$C, dark color), similar to the chocolate fudge ice cream in figure 7.3.

- In the hypoeutectoid region (a plain carbon steel with a composition of less than 0.8 wt. percent carbon), the microstructure would be ferrite and pearlite, as shown in figure 7.5.

- In the hypereutectoid region (a plain carbon steel with a composition of greater than 0.8 wt. percent carbon), the microstructure would be cementite and pearlite.

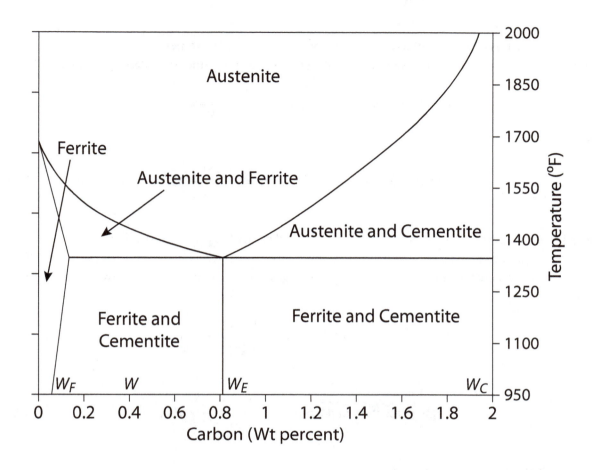

FIGURE 7.4 Iron-Carbon Phase Diagram

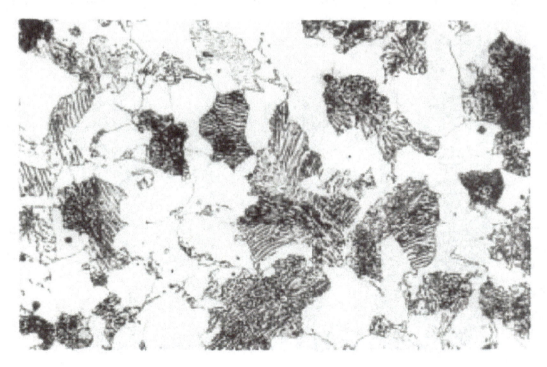

FIGURE 7.5 Microstructure of a Hypoeutectoid Plain Carbon Steel

A phase analysis of a hypoeutectoid plain carbon steel with a W wt. percent carbon at 500°C done with the lever rule would provide weight percentages of ferrite and cementite phases present as follows:

$$\text{Weight percentage of the ferrite} = \frac{W_C - W}{W_C - W_F} \times 100, \tag{7.5}$$

and

$$\text{Weight percentage of the cementite} = \frac{W - W_F}{W_C - W_F} \times 100, \tag{7.6}$$

where W_F and W_C are the compositions of the ferrite and cementite phases, respectively.

A microstructure analysis of a hypoeutectoid plain carbon steel with a W wt. percent carbon at 500°C done with the lever rule would provide weight percentages of ferrite and pearlite microstructure present as follows:

$$\text{Weight percentage of the ferrite} = \frac{W_E - W}{W_E - W_F} \times 100, \tag{7.7}$$

and

$$\text{Weight percentage of the pearlite} = \frac{W - W_F}{W_E - W_V} \times 100, \tag{7.8}$$

where W_E is the eutectoid composition.

When a plain carbon steel is quenched from the high temperature austenite phase, a metastable phase, martensite, is formed. Martensite phase is a solid solution of iron and carbon in a body-centered tetragonal structure, and it is an extremely hard phase with a needle-shaped microstructure, as shown in figure 7.6.

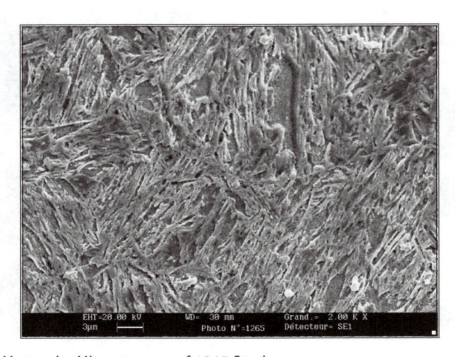

FIGURE 7.6 Martensite Microstructure of 1045 Steel

FIGURE 7.7 Elongated Grains Due to Rolling

Steel alloys from molten stage are cast. After being cast, they are rolled to make different shapes. Rolling elongates the grains in the rolling direction, as shown in figure 7.7, as well as increases strength and hardness (see figure 7.8).

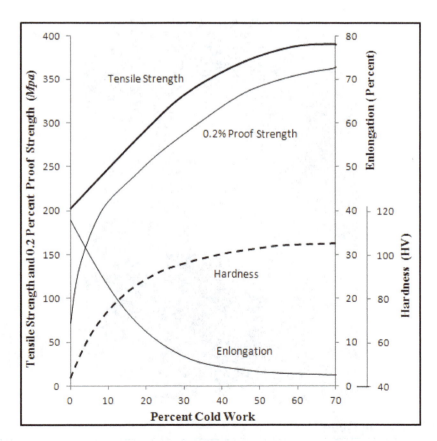

FIGURE 7.8 Variation of Tensile Strength, 0.2 Percent Proof Strength, Hardness, and Elongation as a Function of the Percentage of Cold Work in a Copper Alloy

Test Description

You will be determining the hardness and microstructure of as-received and heat-treated steel samples. Take a picture of the all the equipment to be used along with the test specimens, and include in your report with proper labeling.

Test Procedure

Heat Treatment

- Place two samples (refer to the iron-carbon phase diagram in figure 7.4) in the furnace (Figure 7.9), set at 1600°F, for one hour.

- After one hour, leave one sample to be annealed in the furnace, and let it cool over time (i.e., slow cooling).

- Take out the other sample from the oven and quickly put it in water to quench it.

FIGURE 7.9 An Annealing Oven

Hardness Testing

Measure the Rockwell hardness on all three samples (as-received, annealed, and quenched). Get the mean and standard deviation of HR hardness data, and report this in the data sheet.

Microstructure Measurement

The preparation of a specimen to reveal its microstructure involves the following steps:

- Section or cut the sample into thin pieces, and mount a thin piece on a sample holder. Using rotating wheels, coarse-grind first and then grind on progressively, using finer emery paper, and then polish using alumina powder or diamond paste (Figure 7.10).

- Etch the sample in dilute acid, 2 percent nital for steel (2 vol. percent of nitric acid in ethanol), then wash in alcohol and dry.

- Check the sample under an optical microscope (Figure 7.11) to determine whether it needs more etching. Do not immerse the sample for a long time after the first try to avoid over-etching.

- Examine the sample under the optical microscope. Take photos of all the interesting observations using different magnifications. Typical magnifications used in optical microscope are between 50× and 1000×.

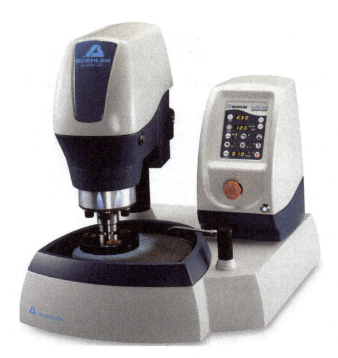

FIGURE 7.10 Grinder and Polisher. Copyright © by Buehler. Reprinted with permission.

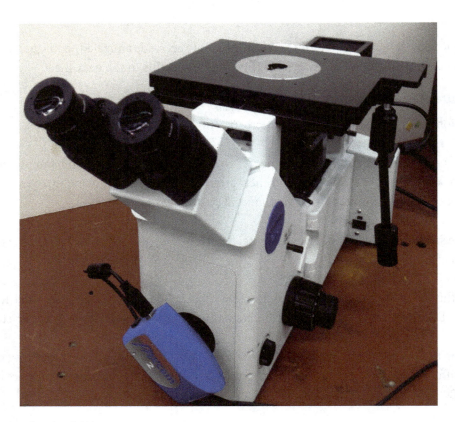

FIGURE 7.11 An Optical Microscope

Materials/Mechanics Laboratory
HTQMA Experiment

HTQMA Hardness Test Data Sheet (Should Be Included in the Report's Appendix)

Specimen and Rockwell Scale	Specimen Composition	HR Reading #1	HR Reading #2	HR Reading #3	Ave.	Std. Dev.
As-Received HRB						
Annealed HRB						
Quenched HRC						

Table 7.1 HR Hardness Values of the Specimens

Specimen and Rockwell Scale	Specimen Composition	Ave. Hardness	Std. Dev.
As-Received HRB			
Annealed HRB			
Quenched HRC			

Results

Hardness

Report hardness values of all three samples in table 7.1.

Get the published data of hardness values for this material, and do a percentage of difference analysis (note that there is no error here; with different heat treatment, the hardness would be changing). Make two column charts to show the comparison, as follows (figures 7.12 and 7.13):

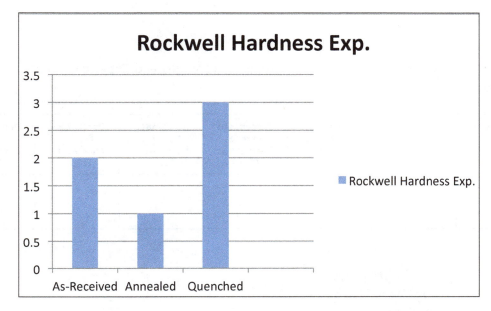

FIGURE 7.12 Example to Display the Rockwell Hardness of Three Samples

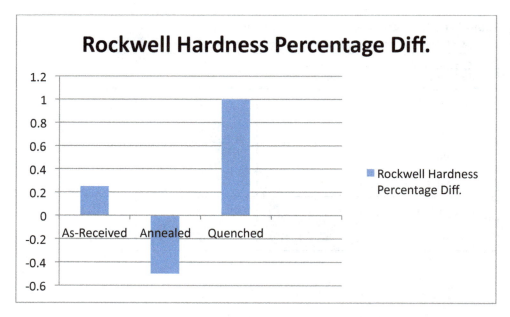

FIGURE 7.13 Example to Display Percentage of Differences of Rockwell Hardness

Microstructure

Include optical images of all three specimens. Using the lever rule, calculate the amount of phases and microstructures and include this in table 7.2. Also, measure the amount of microstructure from the optical image of your sample. Report the experimental values in table 7.2.

Show the experimental and calculated microstructures in a column chart, as shown in figure 7.14. Calculate the percentage of difference of experimental and calculated microstructures.

Table 7.2 Calculated Phases and Microstructures Percentages and Experimental Microstructure Percentages

Spec.	Comp.	Percentage Ferrite Phase (Cal.)	Percentage Cementite Phase (Cal.)	Percentage Ferrite Microstruc. (Cal.)	Percentage Pearlite Microstruc. (Cal.)	Percentage Ferrite Microstruc. (Exp.)	Percentage Pearlite Microstruc. (Exp.)

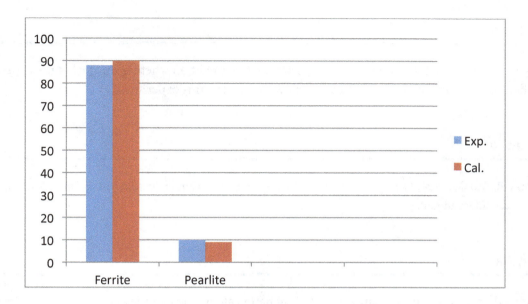

FIGURE 7.14 Example of a Column Chart to Display Experimental and Calculated Values of Microstructures

Discussion

Hardness

Compare the published data for hardness with that of the three samples you have. Comment on the differences in hardness values. It is anticipated that the hardness of the as-received sample (cast and rolled) would be greater than that of the annealed sample. The quenched sample will have the highest hardness.

Microstructure

Discuss all the microstructures obtained. As-received and annealed samples will have pearlite microstructure structure (dark area, lamellar structure, or alternate layers of ferrite [white] and cementite [black]), surrounded by ferrite grain (white area), as shown in figure 7.5. The as-received sample will have elongated grains, as shown in figure 7.7. The quenched sample will have only one phase (martensite) and a needle-shaped microstructure, as shown in figure 7.6. Comment on any discrepancies between the calculated and measured amount of microstructures and percentage differences.

Discuss the correlations between hardness values, heat treatments, and microstructures.

Conclusions

Make your overall conclusions about this experiment. Comment on whether the goal of the experiment is satisfied or not. Give recommendations for how to improve this experiment.

References

[1] Charlie R. Brooks, 1996. *Principles of the Heat Treatment of Plain Carbon and Low Alloy Steels*. Materials Park, OH: ASM International.

Appendix

Present all data sheets and calculations, including hand calculations, in the appendix.

In-Class Problems—HTQMA Experiment

1. A copper wire is cold-drawn from a 2.80 mm to a 2.45 mm diameter. (A) Calculate the percentage of cold work that the wire undergoes. (B) Estimate the wire's tensile strength, 0.2 percent proof strength, hardness, and percentage of elongation.

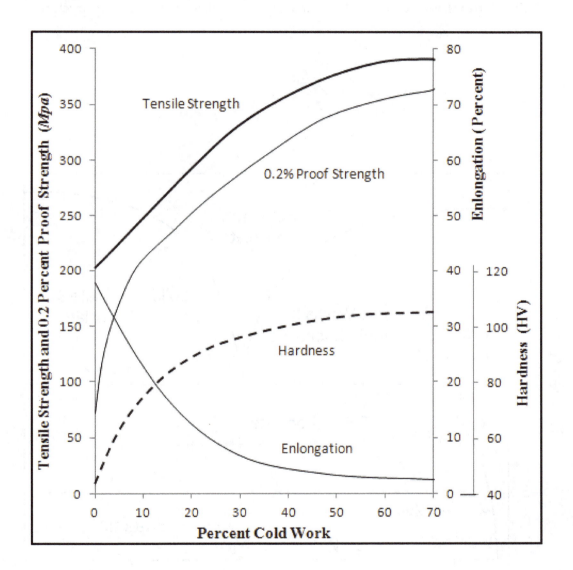

Using the vanilla-chocolate ice cream phase diagram below, answer the following questions:

2. What is the eutectic composition of the vanilla-chocolate ice cream?

3. What are the phases and microstructures present in a 17 percent vanilla-chocolate ice cream at −5°C? Using the lever rule, calculate the amount of each phase and each microstructure.

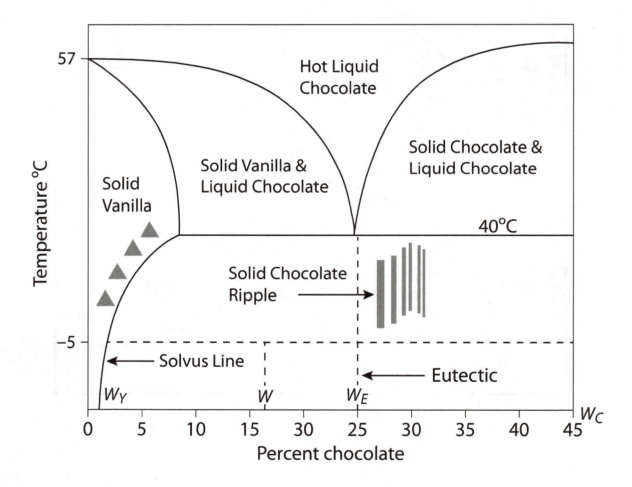

Using the iron-carbon phase diagram below, answer the following questions:

4. What is point *A*? What are the temperature and composition at point *A*?

5. What are the phases present in a 1040 steel at 600°C? Calculate the percentage of each phase present, using the lever rule.

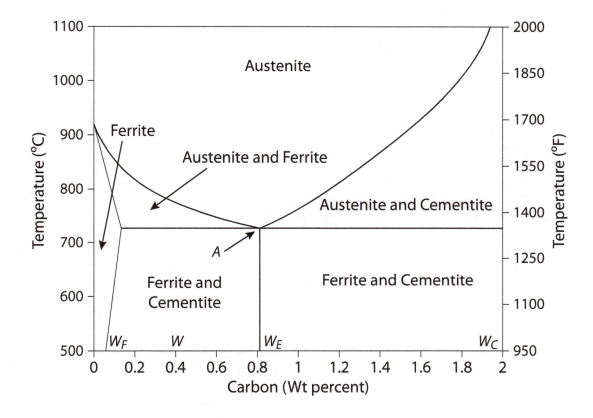

6. What are the microstructures present in a 1040 steel at 600°C? Calculate the percentage of each microstructure present, using the lever rule.

Chapter 8

Torsion Experiment

Introduction

In this experiment we will determine the effects of applying a torsional load to a circular section, such as a circular rod or a tube. There are numerous examples of applications of structural principles of torsion in mechanical, aeronautical, and civil engineering. Materials in the manufacturing industry, such as metal fasteners and beams, are subject to torsion testing to determine their strength under torsion.

The following are some applications: A shaft is a mechanical component for transmitting torque and rotation. Drive shafts (see figure 8.1) are carriers of torque, and they are subject to torsion and shear

FIGURE 8.1 Drive Shaft with Universal Joints at Each End and a Spline in the Center
Copyright © IP83 (CC BY-SA 3.0) at http://en.wikipedia.org/wiki/File:Cardan_Shaft.jpg.

FIGURE 8.2 A Boeing 737 Airliner
Source: http://en.wikipedia.org/wiki/File:Tarom.b737-700.yr-bgg.arp.jpg. Copyright in the Public Domain.

stress. Wings of an aircraft (see figure 8.2) are subjected to bending and shear stresses. Include examples of structures subject to shear stress and torsion in your report.

The objectives of this experiment are to understand the dependence of angular deflection on the applied torque and length of a structure subjected to torsion, and to determine the shear modulus.

It is expected that the angular deflection increases linearly with the applied torque and length. The shear modulus of steel will be greater than that of brass.

Theory [1]

Figure 8.3 shows a body subjected to a shear force, F, over the surface area, A. The shear stress, τ, is given as

$$\tau = \frac{F}{A}. \tag{8.1}$$

The shear strain, γ, is given as

$$\gamma = \frac{\Delta x}{L} = \tan\theta, \tag{8.2}$$

where Δx is the amount of shear displacement and L is the distance over which the shear stress acts. Then, the shear modulus, G, in the elastic region is given as

$$G = \frac{\tau}{\gamma}. \tag{8.3}$$

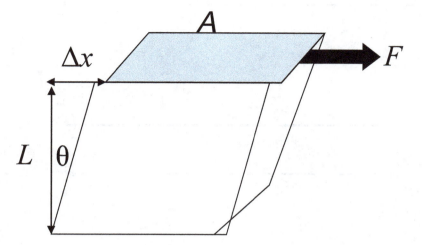

FIGURE 8.3 A Body Subjected to a Shear Force, *F*

Torsion is the twisting of an object due to an applied torque. Considering that a torque is applied to a circular rod, the shear stress is related to the applied torque, *T*, through the following equation:

$$\tau = \frac{TR}{J},$$ (8.4)

where *R* is the radius of the rod, and *J* is the polar moment of inertia of the cross-sectional area and is given as

$$J = \frac{\pi D^4}{32},$$ (8.5)

where *D* is the diameter of the rod.

Figure 8.4 demonstrates angle of twist (angular deflection) under torsion. The shear strain of a rod subjected to torsion can be expressed as

$$\gamma = \frac{R\theta}{L},$$ (8.6)

where *R* and *L* are the radius and length of the rod, respectively, and θ is the angle of twist. The shear strain can also be expressed as

$$\gamma = \frac{\tau}{G} = \frac{TR}{JG}$$ (8.7)

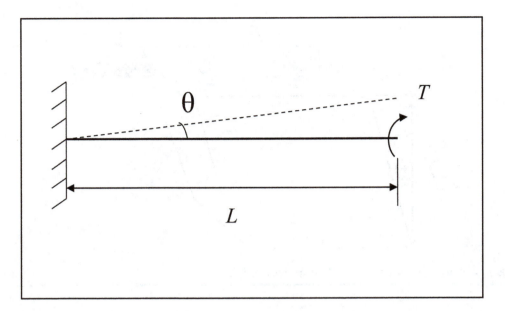

FIGURE 8.4 Demonstration of the angle of Twist and Shearing Strain

From equations 8.6 and 8.7 we can obtain the relation between the angle of twist and torque as follows:

$$\theta = \frac{\gamma L}{R} = \frac{TL}{JG} \tag{8.8}$$

From equation 8.8 we can conclude that the angle of twist (i.e., angle of deflection) depends on the

- Torque linearly; higher torque produces larger angle of twist.

- Length of the rod linearly; there will be higher angle of twist for longer rods or shafts.

- Polar moment of inertia; the angle of twist decreases with the increase in polar moment of inertia.

- Shear modulus; the angle of twist decreases with the increase in shear modulus.

From equation 8.8, shear modulus, G is given as

$$G = \frac{TL}{J\theta} \tag{8.9}$$

Test Description [2]

The experimental setup for torsion testing of circular sections is shown in figure 8.5. It consists of a backboard with chucks at each end. The chucks are for gripping the test specimens. Two circular rods having the same cross-section geometry and the same dimensions will be used, but one will be made of steel, and the other will be made of brass. For the third experiment, a brass rod and a tube will be used. Mount the rod or tube as shown in figure 8.5. Before you start the experiment, take a picture of your test apparatus and samples, and include this in your report. Make sure to label every part of the test apparatus. Three experiments are performed, as follows:

Experiment 1

Keep the length of the rod constant, vary the applied torque, and measure the angular deflection each time, for both brass and steel rods..

Experiment 2

Keep the torque constant, vary the length of the brass rod only, and measure angular defection each time.

FIGURE 8.5 Test Setup for the Torsion Experiment

Experiment 3

Keep the length constant, vary the applied torque, and measure angular deflection of a brass rod and a tube with the same nominal diameter.

Test Procedure

- Visually inspect all parts.

- Check the electrical connections.

- Place the assembled test frame on a workbench.

- Make sure the digital force display is on. Connect the mini DIN from "Force Input 1" on the socket marked "Force Output" to the right underside of the backboard.

- Before you start the experiment, zero the force meter using the dial. Gently apply a small torque to the left-hand chuck and release.

Materials/Mechanics Laboratory
Torsion Experiment

Experiment 1

Torsion Experiment Data Sheet (Should Be Included in the Report's Appendix)

Steel Rod:
Dimensions: Diameter =   Length =   Polar Moment of Inertia (J) =
Torque = Force (N) × Torque Arm Length (0.05 m)

Force (*N*)	Torque (Nm)	Angular Deflection (°)
0		
1		
2		
3		
4		
5		

Brass Rod:
Dimensions: Diameter =   Length =   Polar Moment of Inertia (J) =
Torque = Force (N) × Torque Arm Length (0.05 m)

Force (*N*)	Torque (Nm)	Angular Deflection (°)
0		
1		
2		
3		
4		
5		

Materials/Mechanics Laboratory
Torsion Experiment

Experiment 2

Torsion Experiment Data Sheet (Should Be Included in the Report's Appendix)

Brass Rod:
(Keep Force Constant) $F = 3N$

Length (m)	Angular Deflection (°)
0.30	
0.35	
0.40	
0.45	
0.50	

Materials/Mechanics Laboratory
Torsion Experiment

Experiment 3

Torsion Experiment Data Sheet (Should Be Included in the Report's Appendix)

Brass Rod:
Dimensions: Diameter =   Polar Moment of Inertia (J) =
Torque = Force (N) × Torque Arm Length (0.05 m)
Brass Tube:
Dimensions: Outer Diameter =   Polar Moment of Inertia (J) =
Torque = Force (N) × Torque Arm Length (0.05 m)

Force (N)	Torque (Nm)	Rod Angular Deflection (°)	Tube Angular Deflection (°)
0			
1			
2			
3			
4			
5			

Results

Experiment 1

Include torque and angular deflection from the data sheets in tables 8.1 and 8.2 for steel and brass rods, respectively. Calculate TL and $J\theta$, and include them in tables 8.1 and 8.2. Plot angular deflection versus torque for both rods in one graph to compare. Plot TL versus $J\theta$ for both rods in one graph. Calculate the shear modulus, G, from the slope of the TL versus $J\theta$ for both rods (recall equation 8.9). Calculate the percentage of difference between the experimental and published shear modulus values. Report these data in table 8.3.

Table 8.1 Torque, Angular Deflection, TL, and $J\theta$ for the Steel Rod

Torque (Nm)	Angular Deflection, θ (rad)	TL (Nm²)	$J\theta$ (Nm⁴)

Table 8.2 Torque, Angular Deflection, TL, and $J\theta$ for the Brass Rod

Torque (Nm)	Angular Deflection, θ (rad)	TL (Nm²)	$J\theta$ (Nm⁴)

Table 8.3 Published and Experimental Shear Modulus and Percentage of Error for Both Rods

Beam	Shear Modulus (GPa)		Percentage of Difference
	Experimental	Published	
Steel			
Brass			

Experiment 2

Plot angular deflection versus length for the brass rod.

Experiment 2

Plot angular deflection versus torque for the brass rod and tube in one graph. Make the following table (table 8.4):

Table 8.4 Diameter, Polar Moment of Inertia, and Percentage of Difference in Polar Moment of Inertia for the Brass Rod and Tube

	Diameter (m)	J (Nm⁴)	Percentage of Difference in J
Rod			
Tube	(outer)		

Discussion

Experiment 1

Discuss all the results obtained. Mention that the angular deflection is linearly proportional to the applied torque. Discuss which material has a higher shear modulus, and give your reasons. Mention for which applications you need to minimize the torque (e.g., a drive shaft) and in which applications you need to maximize the torque (e.g., a torque wrench).

Compare the experimental shear modulus with the published one and discuss the percentage of difference. If there is a discrepancy, explain why this exists.

Experiment 2

Mention that the angular deflection is linearly proportional to length. Explain in which applications longer sections are undesirable. After inspecting equation 8.8, if you can't change the length, mention what you can change to minimize the angular deflection.

Experiment 3

Mention that the polar moment of inertia is the same for the tube and the rod with the same nominal diameter, and the missing material at the center of the tube has no effect on the polar moment of inertia. Recall equation 8.8 and mention that the angular deflections of the tube and the rod are the same because T, J, and G are the same.

Conclusions

Make your overall conclusions about this experiment. Comment on what you liked about this experiment and what you did not like. Give recommendations for how to improve this experiment.

References

[1] ASTM Standard E 143, *Standard Test Methods for Shear Modulus at Room Temperature*. West Conshohocken, PA: ASTM International.

[2] *Torsion of Circular Sections: Lecturer Guide*. Nottingham, England: TQ Education and Training Ltd.

Appendix

Present all data sheets and calculations, including hand calculations, in the appendix.

Name: Lab Section:

In-Class Problems—Torsion Experiment

1. The shear modulus of materials is a ratio of the applied _____ to _____.

2. How does the angle of twist (i.e., angle of deflection) depend on (A) applied torque, (B) length, (C) polar moment of inertia, and (D) shear modulus?

3. Calculate the shear modulus when the angular deflection of a tube (of outer diameter 3.2 mm and length 500 mm) is 1.5° under an applied torque of 0.005 Nm.

Chapter 9

Measurement of Strain

Introduction

A strain gage is a device used to measure the strain of an object. Strain gages are used to measure stresses arising from static or dynamic loads from internal or external sources. There are numerous applications of strain gages, some of which are shown below figures 9.1 and 9.2).

FIGURE 9.1 Load Spectrum Measured with Strain Gages at the Wheel Set of a Train Vehicle
Source: http://www.sensortelemetrie.de/en/applications/train/load-spectrum-measurement-whell-load.html. Copyright © by MANNER Sensortelemtry. Reprinted with permission.

FIGURE 9.2 Torque Measurement at a Wind Power Station
Source: http://www.sensortelemetrie.de/en/applikationen/windkraft.html. Copyright © by MANNER Sensortelemtry. Reprinted with permission.

The objectives of the strain experiment are to determine material properties: Poisson's ratio, v, and material stiffness (Young's modulus), E, of a beam, and to understand the usage of strain gages and the Wheatstone bridge. A Wheatstone bridge is used to quantify the resistance changes within the strain gages.

In this experiment we will use a steel beam. The material stiffness (Young's modulus) of any kind of steel (soft steel, structural steel, machine steel, hard steel, or spring steel) has about the same value, generally accepted to be $E = 207$ GPa. This mechanical property of steel is independent of the dimension of the beam, heat treatment, and composition. Only extreme temperature can alter its value. Our experiment is conducted at normal room temperature. Thus, our obtained values of E and v are not going to be affected by temperature.

Theory [1]

The resistance, R, of a wire of length L and cross-sectional area A (see figure 9.3) is given as:

$$R = \frac{\rho L}{A},$$ (9.1)

where ρ is the resistivity of the wire.

Under a tensile force along the length direction, the length changes to $L + \Delta L$, while the cross-sectional area changes to $A - \Delta A$; thus, the resistance changes to $R + \Delta R$, as follows:

$$R + \Delta R = \frac{\rho (L + \Delta L)}{A - \Delta A}$$ (9.2)

A unidirectional stress along the x-direction in a solid body, as shown in figure 9.4, causes a tensile strain, ε_x, along the same direction. Corresponding transverse strains along the y and z directions, ε_y and ε_z, are given as

$$\varepsilon_y = \varepsilon_z = - \text{v} \, \varepsilon_x,$$ (9.3)

where v is the Poisson's ratio.

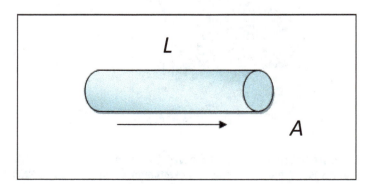

FIGURE 9.3 A Wire Carrying an Electric Current

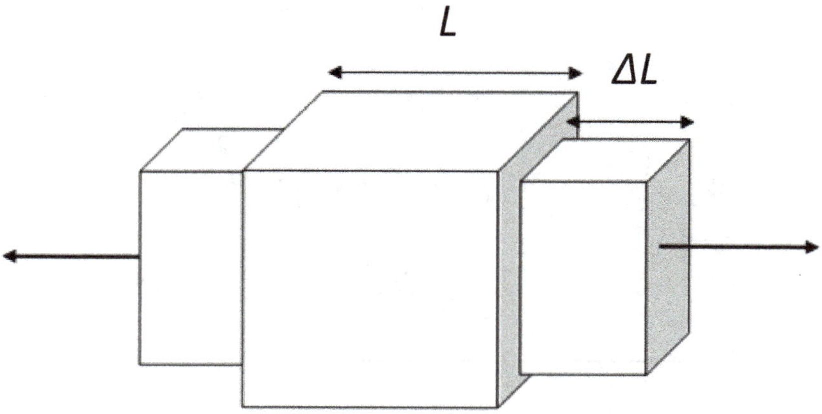

FIGURE 9.4 A Cubic Body under Tensile Stress in the *x* Direction; It Extends Along the *x* Direction while It Contracts along the *y* and *z* Directions, Showing Poisson's Effect

Considering a rectangular wire of length *L*, width *W*, and height *H*, the increase in the resistance of the wire due to a tensile force along the length direction will be

$$R + \Delta R = \frac{\rho\,(L + \varepsilon_x L)}{(W + \varepsilon_y W)(H + \varepsilon_z H)},$$ (9.4)

where the change in length, width, and height, ΔL, ΔW, and ΔH, are given as

$$\frac{\Delta L}{L} = \varepsilon_x, \quad \frac{\Delta W}{W} = \varepsilon_y, \quad \text{and} \quad \frac{\Delta H}{H} = \varepsilon_z.$$ (9.5)

A strain gage is a device whose electrical resistance varies in proportion to the amount of strain in the device. The gage factor (*GF*) is used to express the sensitivity of the strain gage and is defined as the ratio of the fractional change in the resistance to the fractional change in the length, or ε_x:

$$GF = \frac{\Delta R/R}{\Delta L/L} = \frac{\Delta R/R}{\varepsilon_x}.$$ (9.6)

The gage factor for metallic strain gages is typically around 2.

For all practical purposes, strain measurements involve a very small quantity of strain, which results in a very small change in resistance. Such a small change in resistance can be measured using a Wheatstone bridge, as shown in figure 9.5. It consists of four resistive arms of resistances, R_1, R_2, R_3 and R_4, respectively. V_o is the output voltage.

V_{ex}, the excitation voltage, is given as

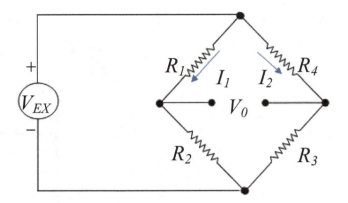

FIGURE 9. 5 A Wheatstone Bridge

$$V_{ex} = I_1(R_1 + R_2) = I_2(R_3 + R_4),$$ (9.7)

where I_1 and I_2 are the currents, as shown in figure 9.5, and, thus,

$$I_1 = \frac{V_{ex}}{R_1 + R_2} \text{ and } I_2 = \frac{V_{ex}}{R_3 + R_4}.$$ (9.8)

The output voltage, V_o, is given by

$$V_o = I_2 R_4 - I_1 R_1.$$ (9.9)

For the bridge to be balanced, V_o needs to be zero, which leads to

$$I_2 R_4 = I_1 R_1 \text{ or } \frac{R_4}{R_3 + R_4} = \frac{R_1}{R_1 + R_2} \text{ or } \frac{R_1}{R_2} = \frac{R_4}{R_3}$$ (9.10)

Any change in resistance in any arm of the bridge results in a nonzero output voltage.

Figures 9.6 and 9.7 describe the experiment to be undertaken. A cantilever beam is bent by placing a load at the free end. Two strain gages are mounted on the top surface of the beam, along the length and width directions. The strain gage along the length direction measures the tensile strain, while the strain gage along the width direction measures the compressive strain. Similarly, two strain gages are mounted on the bottom surface; in this case, the strain gage along the length direction measures the compressive strain, and the strain gage along the width direction measures the tensile strain.

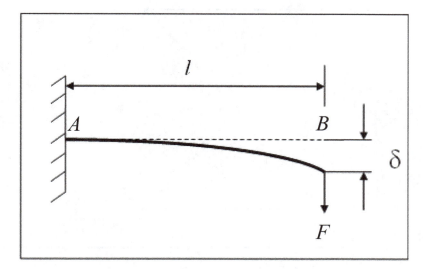

FIGURE 9.6 A Cantilever Beam of Length *l*, Clamped at One End and Loaded at the Other End

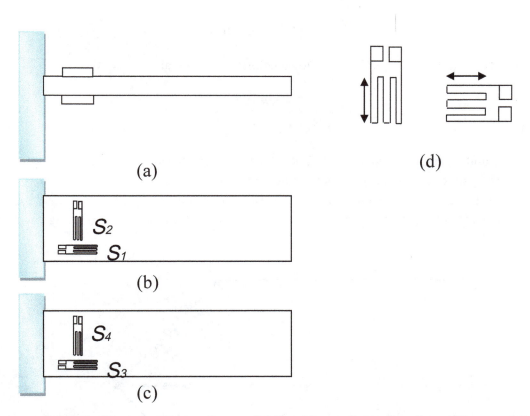

FIGURE 9.7 (a) Unloaded Cantilever Beam; (b) Top View of the Cantilever Beam with Strain Gages Mounted in the Longitudinal (S_1) and Transverse (S_2) Directions; (c) Bottom View of the Cantilever Beam with Strain Gages Mounted in the Longitudinal (S_3) and Transverse (S_4) Directions; (d) Strain Gages

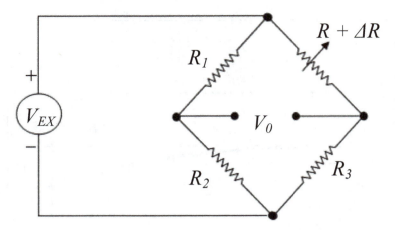

FIGURE 9.8 Quarter-Bridge Circuit

Three different experiments will be performed; these are quarter-bridge, half-bridge, and full-bridge circuit experiments. In the quarter-bridge circuit shown in figure 9.8, R_4 is replaced with the tensile strain gage, S_1, of resistance $R + \Delta R$. The nonzero output voltage, V_o, can be obtained as

$$V_o = I_2(R + \Delta R) - I_1 R_1. \tag{9.11}$$

Substituting I_1 and I_2 from equation 9.8 into equation 9.11, we obtain

$$V_o = V_{ex} \left(\frac{R + \Delta R}{R_3 + R + \Delta R} - \frac{R_1}{R_1 + R_2} \right). \tag{9.12}$$

Measuring V_{ex} and V_o, and using equation 9.12, $R + \Delta R$ can be calculated, and, hence, using equation 9.6, corresponding strain can be obtained if the gage factor is known.

Using a half-bridge circuit as shown in figure 9.9, one can connect both longitudinal and transverse strain gages, S_1 and S_2, by replacing R_4 and R_3, respectively, to obtain the Poisson's ratio. Using a half-bridge

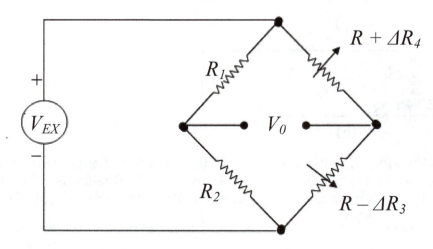

FIGURE 9.9 Half-Bridge Circuit

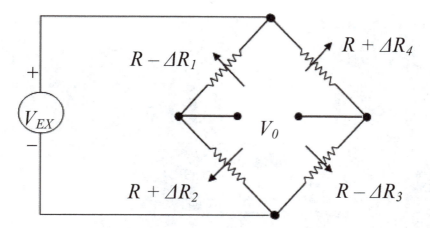

FIGURE 9.10 Full-Bridge Circuit

circuit rather than a quarter-bridge also doubles the sensitivity of the bridge. For a half-bridge circuit, equation 9.12 becomes

$$V_o = V_{ex} \left(\frac{R + \Delta R_4}{R - \Delta R_3 + R + \Delta R_4} - \frac{R_1}{R_1 + R_2} \right),$$
(9.13)

where $R + \Delta R_4$ and $R - \Delta R_3$ are the resistances of S_1 and S_2, respectively.

Finally, one can further increase the sensitivity of the circuit by replacing all four arms of the bridge with active strain gages in a full-bridge configuration, as shown in figure 9.10. R_4 and R_3 are replaced by strain gages S_1 and S_2, and R_1 and R_2 are replaced by strain gages S_3 and S_4 in figure 9.7.

$$V_o = V_{ex} \left(\frac{R + \Delta R_4}{R - \Delta R_3 + R + \Delta R_4} - \frac{R - \Delta R_1}{R - \Delta R_1 + R + \Delta R_2} \right),$$
(9.14)

where $R + \Delta R_2$, $R - \Delta R_1$, $R + \Delta R_4$, and $R - \Delta R_3$ are the resistances of the strain gages S_4, S_3, S_1, and S_2, respectively.

Test Description

Figure 9.11 shows the apparatus to be used for the strain experiment. A steel cantilever beam with two strain gages mounted on the top and two strain gages mounted on the bottom is used. Figure 9.12 shows a typical strain gage that is used. The metallic strain gage consists of a very fine wire arranged in a grid pattern. The grid is bonded to a thin backing, which is attached directly to the test specimen. Different weights are put on the weight holder; the strain gages are connected via the quarter-bridge, half-bridge, and full-bridge experiments, following the procedure below. The strain value is obtained directly from the Wheatstone bridge. Take a picture of the equipment and test specimen, and include in your report, with proper labeling.

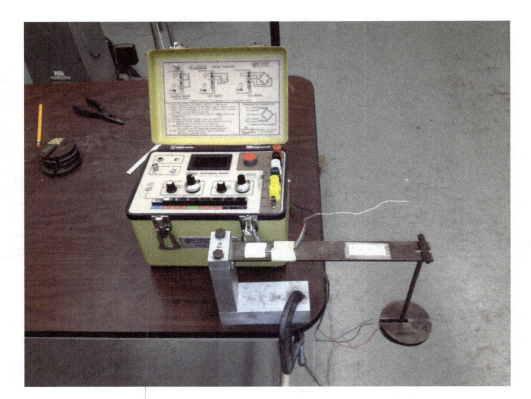

FIGURE 9.11 Strain Experiment Setup

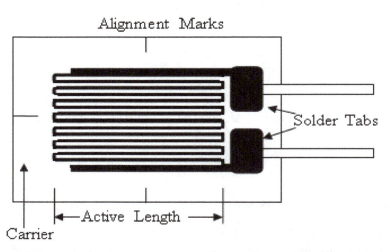

FIGURE 9.12 Bonded Metallic Strain Gage

Test Procedure

- Note down the gage factor and resistance of the strain gage.

- Perform the quarter-, half-, and full-bridge experiments by adding masses.

- Obtain the strain value each time, and note this down in the data sheet.

Materials/Mechanics Laboratory
Strain Experiment

Strain Experiment Data Sheet (Should Be Included in the Report's Appendix)

Experimental Parameters:
Gage Factor: Resistance:
Beam Dimensions
Thickness: Width: Length:
Masses:
A: B: C: D: E:

Indicated Strain Values (Micro Strain):

Mass (g)	Quarter Bridge Axial ($BSF = 1$)	Quarter Bridge Transverse ($BSF = -v$)	Half Bridge Top Gages ($BSF = v + 1$)	Full Bridge ($BSF = 2v + 2$)
A				
A + B				
A + B + C				
A + B + C + D				
A + B + C + D + E				

Note: *BSF* = bridge sensitivity factor.

Results

Plot $\varepsilon_{transverse}$ against ε_{axial}, and calculate ν from the slope of the plot.

Calculate actual strain from indicated strain as

$$\varepsilon_{actual} = \frac{\varepsilon_{indicated}}{BSF}, \tag{9.15}$$

where BSF is the bridge sensitivity factor.

Calculate the bending stress, σ, of the beam using the following equation:

$$\sigma = \frac{Mc}{I}, \tag{9.16}$$

where M is the moment of the beam, c is the distance of the central axis of the beam from the top surface, and I is the area moment of inertia of the beam. The expressions for M, c, and I are as follows:

$$M = mgl, \tag{9.17}$$

$$c = \frac{h}{2}, \tag{9.18}$$

and

$$I = \frac{bh^3}{12}, \tag{9.19}$$

where m is the mass, and l, h, and b are the length, thickness, and width of the beam, respectively.

Substituting the values of M, c, and I in equation 9.16 with equations 9.17 through 9.19, respectively, bending stress can be obtained as

$$\sigma = \frac{6mgl}{bh^2}. \tag{9.20}$$

Construct table 9.1. Plot stress versus actual strain for all four cases. Determine Young's modulus, E, from the slope from all four plots. Calculate the percentage of difference of ν and E between the experimental and published data, and include in tables 9.2 and 9.3 respectively.

Table 9.1 Bending Stress and Strain Values for Different Bridge Conditions

Mass (g)	σ (MPa)	ε_{actual} (m/m) Quarter Bridge Axial	ε_{actual} (m/m) Quarter Bridge Transverse	ε_{actual} (m/m) Half Bridge Top Gages	ε_{actual} (m/m) Full Bridge

Table 9.2 Published and Experimental Poisson's Ratio and Percentage of Difference

$\nu_{Exp.}$	$\nu_{Publ.}$	% Difference

Table 9.3 Published and Experimental Elastic Modulus and Percentages of Difference

$E_{Publ.}$	$E_{Exp.}$ Quarter Bridge Axial	% Diff.	$E_{Exp.}$ Quarter Bridge Trans.	% Diff.	$E_{Exp.}$ Half Bridge	% Diff.	$E_{Exp.}$ Full Bridge	% Diff.

Discussion

Compare the published Poisson's ratio of steel with the experimental one. Discuss the percentage of difference.

Compare the published material stiffness (Young's modulus, E) of steel with the experimental one separately for all four cases. Comment and explain on the accuracy of all four cases for E value. Discuss the percentage of difference.

Conclusions

Make your overall conclusions about this experiment. Comment on what you liked about this experiment and what you did not like. Give recommendations for how to improve this experiment.

References

Richard L. Hannah and Stuart E. Reed, Society for Experimental Mechanics, London: Chapman & Hall; Bethel 1994.

Appendix

Present all data sheets and calculations, including hand calculations, in the appendix.

Name: **Lab Section:**

In-Class Problem—Strain Experiment

1. Calculate Poisson's ratio of a material when longitudinal strain is 0.12 m/m and transverse strain is −0.0396 m/m.

2. If a rectangular wire is 10 cm in length, 2 cm in width, and 2 cm in height, and is made of the material in problem 1 and undergoes the same longitudinal strain as in problem 1, what will be the value of change in resistance, ΔR? Consider resistivity, $\rho = 0.5$ ohm.cm.

3. Calculate changes in resistance due to longitudinal and transverse strains when the longitudinal strain is 0.12 m/m and transverse strain is −0.0396 m/m for a material. The gage factor is 2, $R = 100$ Ohms.

4. Calculate V_o when $V_{ex} = 6$ volts and you are using a half-bridge circuit to measure the strains described in problem 3. R_0, R_1, and $R_2 = 100$ Ohms.